Green Start
Simple Steps to an Eco-Friendly Life

"Green Start: Simple Steps to an Eco-Friendly Life"

Are you ready to make a positive impact on the planet but don't know where to start? *Green Start: Simple Steps to an Eco-Friendly Life* is your essential guide to living sustainably in today's world. Perfect for beginners, this book breaks down the overwhelming task of "going green" into manageable, practical steps that anyone can follow.

From reducing household waste and choosing sustainable food options to conserving energy and water, *Green Start* offers actionable advice on how to build eco-friendly habits that last. You'll discover the core principles of sustainability, learn how small lifestyle changes can make a big impact, and gain tips on everything from eco-conscious fashion to green transportation.

Key topics include:

- Understanding the environmental footprint of daily actions.
- Simple strategies for waste reduction using the 5Rs (Refuse, Reduce, Reuse, Recycle, Rot).
- Mindful food shopping, reducing food waste, and eating local and seasonal.
- Easy ways to save energy and water at home.
- Sustainable fashion choices and reducing consumption.
- Creating green spaces with gardening and composting.

Each chapter is designed to help you build long-term sustainable habits, with insights on how to involve family and friends, stay motivated, and even deal with eco-anxiety. Whether you're just starting out or looking to deepen your eco-knowledge, this book will inspire you to take meaningful steps toward a greener, healthier lifestyle.

With *Green Start*, you'll not only help protect the planet, but also enjoy personal benefits like saving money, improving your health, and living a more intentional life. Let's make sustainable living accessible for everyone, one small step at a time!

"Green Start: Simple Steps to an Eco-Friendly Life"
Copyright 2024. All rights reserved.

Introduction

1. **Why Go Green?** Understanding the importance of eco-friendly living.
2. **Small Changes, Big Impact.** How small actions can make a significant difference.
3. **How to Use This Book.** A roadmap for gradual, sustainable changes.

Chapter 1: Getting Started with Sustainable Living

1. **What Does "Eco-Friendly" Really Mean?** Defining sustainability and its key principles.
2. **Assessing Your Current Lifestyle.** Identifying areas for improvement.
3. **Setting Realistic Goals.** How to create manageable, eco-friendly habits.

Chapter 2: Waste Reduction Basics

1. **Understanding Waste.** The life cycle of waste and its environmental impact.
2. **The 5Rs: Refuse, Reduce, Reuse, Recycle, Rot** Core strategies for reducing waste.
3. **Practical Ways to Reduce Household Waste.** Tips for minimizing waste in the kitchen, bathroom, and daily life.
4. **Switching to Reusables.** Alternatives to single-use products.

Chapter 3: Sustainable Food Choices

1. **Why Food Choices Matter.** The environmental impact of our diet.
2. **Shopping Mindfully.** Tips for reducing food packaging and buying in bulk.
3. **Eating Seasonal and Local.** Benefits of supporting local farmers and eating with the seasons.
4. **Reducing Food Waste.** How to store food properly and use leftovers creatively.

Chapter 4: Energy Conservation at Home

1. **Energy Use and Its Environmental Impact.** How everyday energy use contributes to climate change.
2. **Simple Tips to Save Electricity.** Reducing energy consumption with small actions.
3. **Efficient Heating and Cooling.** How to maintain comfort while conserving energy.
4. **Renewable Energy Options.** Exploring solar, wind, and other renewable energy sources.

Chapter 5: Water Conservation Made Easy

1. **The Importance of Water Conservation.** Understanding water scarcity and its global implications.
2. **How to Reduce Water Use at Home.** Practical steps for saving water in the kitchen, bathroom, and garden.
3. **Collecting and Reusing Water.** Rainwater harvesting and other water-saving techniques.

Chapter 6: Eco-Friendly Transportation

1. **Transportation and Carbon Footprint.** How our travel habits affect the planet.

2. **Choosing Greener Ways to Travel.** Walking, biking, carpooling, and public transportation.
3. **Switching to Electric or Hybrid Vehicles.** The benefits and considerations of eco-friendly cars.
4. **Reducing Travel Emissions.** How to minimize your environmental impact when traveling.

Chapter 7: Sustainable Fashion and Consumption

1. **The Impact of Fast Fashion.** Understanding the environmental cost of clothing production.
2. **Slow Fashion: Quality Over Quantity.** Choosing durable, ethically made clothes.
3. **Secondhand and Upcycling.** How to shop sustainably and give new life to old items.
4. **Minimalism and Mindful Consumption.** Reducing overall consumption and buying with intention.

Chapter 8: Green Cleaning and Home Care

1. **Toxic Chemicals in Household Products.** The dangers of conventional cleaning products.
2. **Making Your Own Eco-Friendly Cleaners.** DIY recipes for non-toxic cleaning solutions.
3. **Sustainable Home Maintenance.** How to keep your home in good shape while reducing your environmental impact.

Chapter 9: Creating Green Spaces

1. **Gardening for Beginners.** How to start your own eco-friendly garden.
2. **Composting Made Simple.** Turning food waste into nutrient-rich compost.
3. **Growing Your Own Food.** The benefits of growing fruits, vegetables, and herbs at home.

Chapter 10: Forming Sustainable Habits

1. **Building Long-Term Eco-Friendly Habits.** Tips for staying motivated and committed.
2. **Involving Family and Friends.** How to inspire others to join your green journey.
3. **Dealing with Eco-Anxiety.** How to stay positive and avoid feeling overwhelmed by environmental issues.

Conclusion

1. **Your Green Journey.** Reflecting on your progress and future goals.
2. **What's Next?** Continuing to learn and grow in your eco-friendly lifestyle.

Appendices

- **Eco-Friendly Product Recommendations.** A list of sustainable brands and products.
- **Checklists for Green Living**
  Printable guides to help you stay on track with your eco-friendly habits.

Introduction

Why Go Green?

In today's world, the effects of human activity on the environment are impossible to ignore. From rising global temperatures to the alarming rate of species extinction, the need for sustainable living has never been more pressing. But what does it really mean to "go green," and why should you make the effort to live more sustainably?

Going green means adopting practices that reduce your environmental footprint—the impact your lifestyle has on the planet's resources and ecosystems. Every day, we make choices that either contribute to the problem or help solve it. Whether it's choosing a product made from recycled materials, reducing energy consumption, or switching to reusable items, each decision plays a role in shaping the future of our planet.

But sustainability isn't just about environmental responsibility. It's also about improving your quality of life. Eco-friendly habits can lead to healthier living by reducing exposure to toxins and pollutants, encouraging a diet that's better for you and the environment, and fostering a deeper connection with the natural world. On a personal level, many eco-friendly practices can also save you money—think of the energy you'll save by switching to efficient appliances or the money you'll keep by reducing food waste.

At its core, sustainability is about balance: meeting our current needs without compromising the ability of future generations to meet theirs. This book is designed to help you find that balance, providing you with practical steps to incorporate eco-friendly habits into your daily life—gradually and without overwhelm.

Small Changes, Big Impact

One of the biggest misconceptions about sustainable living is that it requires drastic, immediate changes. Many people feel paralyzed by the idea that they need to overhaul their entire lifestyle in order to make a difference. But the truth is, even small, consistent changes can have a big impact over time.

This book is built on the idea that making one small, positive change can lead to another, and another, until those individual actions add up to a major lifestyle shift. It's not about being perfect; it's about being conscious and making better choices when you can. Whether it's opting for reusable bags, conserving water at home, or buying local produce, each small step you take brings you closer to living in harmony with the planet.

Moreover, small changes are often more sustainable in the long run. When we try to do too much at once, it's easy to get overwhelmed and give up. By starting with simple, manageable actions, you build confidence and momentum that can carry you toward bigger, more impactful changes.

How to Use This Book

"Green Start: Simple Steps to an Eco-Friendly Life" is designed to guide you step by step through the process of adopting sustainable habits. Each chapter focuses on a specific area of life where you can make eco-friendly improvements—from reducing waste and making sustainable food choices to conserving energy and water at home.

The book is structured to offer practical, achievable steps that you can integrate into your daily routine. You don't need to do everything at once. In fact, this book encourages you to take it slow—start with the changes that feel most manageable, then build on them over time.

Each chapter ends with a checklist of simple actions you can take right away. Use these checklists as guides to track your progress. By the end of the book, you'll have developed a personalized plan for living a more eco-friendly life—one that fits your lifestyle and feels sustainable for the long term.

Remember, the goal isn't perfection. It's progress. Every small step you take toward a greener life is a step in the right direction for you, your family, and the planet. Let's get started!

Chapter 1: Getting Started with Sustainable Living

What Does "Eco-Friendly" Really Mean?

At its core, "eco-friendly" refers to practices and choices that are designed to have minimal impact on the environment. It's a term often used to describe products, behaviors, or lifestyles that help preserve natural resources and reduce pollution, waste, and energy consumption. However, being eco-friendly is not just about individual actions but about adopting a broader, more conscious approach to the way we live, consume, and interact with the planet.

Breaking Down the Concept of Sustainability

To fully understand what it means to be eco-friendly, it's important to look at the broader concept of sustainability. Sustainability focuses on meeting the needs of the present without compromising the ability of future generations to meet their own needs. It involves balancing environmental, economic, and social factors to ensure that we maintain the health of the planet while supporting human well-being.

Being eco-friendly is part of this sustainable mindset. It means making choices that conserve resources like water, energy, and raw materials, and that reduce harm to ecosystems. It also means thinking about the long-term impact of our decisions—whether it's the type of food we eat, the transportation we use, or the waste we produce.

Key Principles of Eco-riendly Living

Here are the key principles that underpin eco-friendly living:

1. **Minimizing Waste**: An eco-friendly lifestyle focuses on reducing the amount of waste you produce. This includes limiting your use of single-use plastics, recycling whenever possible, composting organic waste, and finding ways to reuse or repurpose items.
2. **Conserving Resources**: Resources like water, energy, and raw materials are finite, and eco-friendly living aims to use them more efficiently. This can involve turning off lights when not in use, fixing leaks to conserve water, or choosing products made from recycled or sustainably sourced materials.
3. **Reducing Pollution**: Pollution—whether in the form of greenhouse gases, chemical runoff, or plastic waste—has devastating effects on the environment. Eco-friendly living seeks to minimize pollution by

choosing cleaner energy sources, reducing the use of harmful chemicals, and limiting reliance on fossil fuels.
4. **Supporting Sustainable Production**: When possible, being eco-friendly means supporting companies and products that prioritize sustainability. This might involve buying organic produce, choosing fair-trade goods, or supporting local businesses that practice responsible farming or manufacturing.
5. **Prioritizing Renewable Energy**: Shifting from fossil fuels to renewable energy sources like solar, wind, or hydropower is a crucial step toward reducing carbon emissions and slowing climate change. Eco-friendly choices often include finding ways to reduce energy consumption or investing in renewable energy options.

The Role of Individual Actions

It's easy to feel overwhelmed by the scale of environmental issues like climate change or plastic pollution. But individual actions, when adopted widely, can collectively make a big difference. Eco-friendly living isn't about perfection; it's about making thoughtful, informed decisions in your daily life that align with the principles of sustainability.

For example, choosing to bring a reusable bag to the store might seem small, but over time, and if many people adopt the same habit, it can significantly reduce the amount of plastic waste. Similarly, small efforts to conserve water, reduce energy consumption, or avoid fast fashion can add up to big savings in natural resources and help protect the environment.

Eco-friendly living is about recognizing that we all have a role to play in protecting the planet. By making choices that prioritize sustainability, we can reduce our impact on the environment and help create a more resilient, healthier world for future generations.

In summary, living eco-friendly means understanding the relationship between our everyday choices and their environmental impacts. It involves being mindful of the resources we use, the waste we generate, and the pollution we create. Most importantly, it's about committing to making small, consistent changes that help us move toward a more sustainable and harmonious way of life with the planet.

Assessing Your Current Lifestyle

Before making any changes toward a more eco-friendly lifestyle, it's important to assess your current habits and identify the areas where you can make the most impact. This self-assessment is a vital first step, as it will help you understand where you are using the most resources, generating waste, or contributing to environmental harm—and where small adjustments can lead to meaningful changes.

Here are some key areas of your life to examine, along with questions to guide your self-assessment:

1. Energy Use at Home

Energy consumption is one of the largest contributors to an individual's carbon footprint. Assessing how you use energy in your home can reveal opportunities for improvement.

- **Heating and Cooling**: Do you often leave the thermostat set too high or low? Can you reduce heating or cooling by using energy-efficient appliances, improving insulation, or using natural ventilation (open windows, fans)?
- **Appliances**: How old and energy-efficient are your appliances (refrigerator, washing machine, dryer, etc.)? Are you using energy-saving settings on your appliances, such as lower temperature wash cycles or air-drying clothes instead of using a dryer?
- **Lighting**: Are you using energy-efficient bulbs (like LED or CFL bulbs)? Do you turn off lights when leaving rooms or when they're not needed?
- **Electronics**: Do you leave devices plugged in when they're not in use? Many electronics continue to draw power when left plugged in, which is known as "phantom" or "standby" power.

Action Steps to Consider:

- Switch to energy-efficient lighting and appliances.
- Use a programmable thermostat to control heating and cooling.
- Unplug devices when not in use or invest in a power strip with an on/off switch.

Water Consumption

Water is a precious and increasingly scarce resource. The average person uses gallons of water every day, often without considering how much they're using or wasting.

- **Water-Efficient Fixtures**: Do you have water-saving faucets, showerheads, and toilets? Are you taking long showers or leaving the tap running while brushing your teeth?

- **Laundry and Dishwashing**: Do you run your washing machine or dishwasher with a full load, or are you washing small loads regularly? Are you using eco-friendly detergents that are biodegradable and free from harmful chemicals?
- **Outdoor Water Use**: Do you use water efficiently in your garden or lawn care? Do you rely on a hose for watering, or do you collect rainwater to use for gardening?

Action Steps to Consider:

- Install water-saving fixtures and appliances.
- Reduce water waste by turning off taps when not in use.
- Consider using a rainwater harvesting system for your garden or landscaping.

3. Waste Generation

Waste is one of the most visible signs of unsustainable living. Many of the things we use on a daily basis are single-use or disposable, ending up in landfills where they can take hundreds of years to decompose.

- **Packaging**: How much of your food, clothing, and products come in excessive or non-recyclable packaging? Do you rely on single-use plastics such as water bottles, bags, or straws?
- **Food Waste**: How much food do you throw away? Do you let produce rot in the fridge, or do you throw out leftovers? Are you able to compost your food scraps instead of sending them to the landfill?
- **Recycling**: Are you familiar with what can and can't be recycled in your area? Are you separating recyclable materials from your regular trash?

Action Steps to Consider:

- Reduce your consumption of packaged and single-use items.
- Start composting food scraps to divert waste from landfills.
- Educate yourself on proper recycling practices and set up a recycling station at home.

4. Transportation Choices

Transportation is a major contributor to greenhouse gas emissions. Assessing your transportation habits can reveal opportunities for reducing your environmental impact.

- **Car Use**: How often do you drive? Is your car fuel-efficient, or does it consume a lot of fuel? Do you drive short distances, when walking or biking might be an option?

- **Public Transportation and Carpooling**: Do you use public transportation, carpool with others, or take advantage of ride-sharing services? Or do you drive alone everywhere you go?
- **Air Travel**: How often do you fly? Air travel has a large carbon footprint, especially for long-distance flights.

Action Steps to Consider:

- Walk, bike, or take public transportation whenever possible.
- Consider using ride-sharing services or carpooling to reduce the number of cars on the road.
- Limit air travel or offset your carbon emissions when you do fly.

5. Consumption and Shopping Habits

The products we buy and how we use them have a significant environmental impact, from the raw materials used in their production to their disposal at the end of their life cycle.

- **Clothing**: Do you frequently buy fast fashion, or do you choose high-quality, long-lasting items? Do you donate or repurpose clothes rather than throwing them away?
- **Food Choices**: Do you buy locally grown, organic, or sustainably sourced foods? Are you aware of the environmental impacts of the meat and dairy you consume?
- **Packaging and Waste**: How often do you purchase products with excessive packaging, or items that you know will only be used once and then discarded?

Action Steps to Consider:

- Buy high-quality, durable goods that are built to last.
- Shop secondhand or exchange items with friends and family.
- Choose plant-based foods more often and support local, sustainable food systems.
- Avoid items with excessive or non-recyclable packaging.

6. Eco-Friendly Habits in the Workplace or School

Your environmental impact extends beyond your home—whether you work in an office, study at school, or have a remote job. How can you bring eco-friendly practices into these spaces?

- **Paper Use**: Are you using paper efficiently? Do you print documents unnecessarily, or do you prefer digital records whenever possible?

- **Energy Efficiency**: Does your workplace have energy-efficient lighting, heating, and cooling? Do you switch off equipment when not in use?
- **Sustainable Practices**: Are there opportunities to encourage eco-friendly practices at work or school, such as recycling programs, energy-saving initiatives, or sustainable commuting options?

Action Steps to Consider:

- Reduce paper waste by going digital and printing only when necessary.
- Encourage your workplace or school to adopt sustainable practices, like recycling or using energy-efficient appliances.
- Turn off lights and electronics when not in use.

Conclusion

By assessing your lifestyle in these key areas, you can begin to identify where changes are needed and where improvements are possible. Remember, this process doesn't need to be overwhelming. Take small steps, set achievable goals, and gradually build up your sustainable habits. Start by making one or two changes, and as you feel more comfortable, expand your efforts to other areas of your life. The key is to make consistent, mindful decisions that reduce your environmental impact and move you toward a more eco-friendly lifestyle.

In the next chapter, we will explore practical ways to reduce waste and start incorporating sustainable habits into your everyday routine.

Setting Realistic Goals: How to Create Manageable, Eco-Friendly Habits

Adopting an eco-friendly lifestyle doesn't happen overnight, and it's easy to feel overwhelmed when thinking about all the changes you could make. The key to successfully transitioning to a more sustainable way of living is to set realistic, achievable goals that you can gradually build upon. By focusing on small, manageable steps, you create eco-friendly habits that become a natural part of your daily routine rather than feeling like burdensome tasks.

Here's a step-by-step guide on how to set realistic goals for creating lasting, eco-friendly habits:

1. Start Small and Be Specific

One of the most effective strategies for setting eco-friendly goals is to start with small, specific actions that are easy to incorporate into your lifestyle.

Vague goals like "become more eco-friendly" can be overwhelming and difficult to track, so break them down into clear, actionable steps.

Examples of small, specific goals:

- "Switch to reusable shopping bags this week."
- "Use energy-efficient light bulbs in the living room by the end of the month."
- "Eat at least one plant-based meal every week for the next month."
- "Switch to a bamboo toothbrush by the end of the month."

These goals are easy to measure and implement, which makes them feel achievable, and they set a positive tone for the bigger changes that will follow.

2. Focus on One Area at a Time

Trying to change everything at once can feel overwhelming and lead to burnout. Instead, focus on one area of your lifestyle to improve, and once you feel comfortable with the new habits, move on to another. Prioritize the areas that will have the biggest impact on your environmental footprint or that feel the easiest to implement.

Example:

- If waste reduction is a priority, start by focusing on eliminating single-use plastics. Replace disposable water bottles with reusable ones, buy products with minimal packaging, or switch to cloth napkins instead of paper towels.
- If energy conservation is important to you, start by reducing your electricity usage. Switch to LED bulbs, unplug devices when not in use, or invest in a programmable thermostat to optimize your home's heating and cooling.

Once you've made consistent progress in one area, move on to another. This will help prevent you from feeling like you have to do everything at once and make the process feel more manageable.

3. Set Achievable Timelines

While it's important to challenge yourself, setting goals with an unrealistic timeline can make you feel frustrated or give up. Instead, create goals with realistic timelines that give you enough time to succeed without feeling rushed.

For example:

- *"In the next three months, reduce my household energy consumption by 10%."*
- *"By the end of the year, switch all of my cleaning products to non-toxic, eco-friendly alternatives."*
- *"In the next month, reduce food waste by meal planning and composting."*

Give yourself the flexibility to adjust your goals if needed, but set a reasonable timeframe that will allow you to build lasting habits.

4. Make It Habitual

One of the most effective ways to create eco-friendly habits is to integrate them into your daily routine so that they feel like second nature. Rather than relying on motivation alone, build systems that make eco-friendly choices easier.

Tips to make eco-friendly habits stick:

- **Set reminders**: Use your phone or sticky notes to remind yourself of the eco-friendly habits you're working on. For example, put a reminder on your fridge to "take reusable bags when shopping."
- **Track your progress**: Keep a log or journal of your eco-friendly actions. This could be as simple as noting when you remembered to bring your reusable cup or when you cooked a meatless meal.
- **Create routines**: Link new habits with existing ones. For instance, if you already have a morning coffee routine, use that time to also remember to take your reusable coffee cup with you.
- **Reward yourself**: Celebrate the small victories! Each time you successfully stick to an eco-friendly habit, take a moment to acknowledge the positive impact you're making. This reinforces your commitment and helps you stay motivated.

5. Be Flexible and Adjust as You Go

Sometimes, things won't go as planned, and that's okay! Life is unpredictable, and sustainable living doesn't require perfection. Don't be discouraged if you slip up—what matters is getting back on track and staying committed to your long-term goals.

For example, if you set a goal to reduce plastic use but find yourself needing to buy a product with plastic packaging one week, that's okay. Reflect on what led to that decision, and adjust your habits accordingly. Maybe the next time you'll bring your own container or find a more sustainable option.

Flexibility and adaptability are key. Sustainable living isn't a one-time project, it's a journey of continuous improvement, and the more forgiving you are with yourself, the more likely you are to stick with it.

6. Make Your Goals Measurable

Measuring your progress helps keep you on track and provides motivation to continue. Setting specific metrics to track will give you a sense of accomplishment and show you just how much impact your actions are having.

Examples of measurable goals:

- "Reduce household waste by 20% in six months."
- "Save 15% on my monthly energy bill by switching to energy-efficient appliances and practices."
- "Buy at least 75% of my produce from local farmers or markets for the next three months."

When you can quantify your progress, it's easier to see how small changes are adding up over time, and that momentum will help you stay motivated to achieve your next set of goals.

7. Celebrate Your Successes

Finally, it's important to celebrate your successes along the way, no matter how small they may seem. Acknowledging your achievements helps reinforce the positive behavior and reminds you of the impact you're making. Every step counts, and it's crucial to recognize the progress you've made.

For example:

- If you successfully go a month without using any plastic bags, give yourself a pat on the back.
- If you reduce food waste by 30%, celebrate with a small reward, like treating yourself to a sustainable product you've been eyeing.

By celebrating these milestones, you'll not only feel proud of your efforts but also build motivation to tackle your next eco-friendly goal.

Conclusion

Setting realistic, manageable goals is the cornerstone of creating lasting eco-friendly habits. Remember to start small, be specific, and focus on one area at a time. Track your progress, adjust when needed, and celebrate your successes along the way. The more you make eco-friendly actions part of your daily routine, the easier it will become to sustain them long-term. Through consistent, gradual changes, you can make a meaningful difference in the world—and inspire others to do the same.

Chapter 2: Waste Reduction Basics

1. Understanding Waste: The Life Cycle of Waste and Its Environmental Impact

Waste is a critical issue in our modern world, and understanding its life cycle and environmental impact is key to reducing it effectively. Waste is more than just something we throw away—it's part of a larger process that begins with the extraction of raw materials and ends with disposal in landfills, incinerators, or recycling facilities. The journey of waste involves many stages, each of which can have significant environmental consequences.

To truly grasp the importance of reducing waste, it's essential to understand how waste impacts the environment at each stage of its life cycle.

The Life Cycle of Waste

The life cycle of waste can be broken down into several stages, from raw material extraction to disposal. Here's a closer look at each stage:

1. **Extraction of Raw Materials**
   Every product we use starts with the extraction of raw materials, such as oil, metal ores, wood, or agricultural products. This stage often involves resource depletion, habitat destruction, and pollution. Mining, logging, and drilling can cause soil erosion, deforestation, water contamination, and loss of biodiversity.

   Example: The production of plastic starts with the extraction of petroleum, a finite resource. This process can contribute to pollution and habitat destruction, as well as carbon emissions.

2. **Manufacturing and Processing**
   After raw materials are extracted, they are processed and manufactured into finished products. This stage requires energy, water, and chemicals, which can all contribute to pollution and environmental degradation. The manufacturing process can produce waste, including chemical runoff, particulate matter, and greenhouse gas emissions.

   Example: The manufacturing of electronics involves mining rare earth metals, using toxic chemicals, and emitting carbon dioxide. These processes can pollute local water sources and ecosystems.

3. **Distribution and Transportation**
   Once products are manufactured, they need to be transported to stores, warehouses, and consumers. This step often requires significant energy and contributes to carbon emissions, as most goods are transported using trucks, ships, or airplanes powered by fossil fuels. Transportation also increases packaging waste to ensure products are protected during transit.

   Example: Imported goods, such as clothing or electronics, may travel thousands of miles to reach consumers, contributing to carbon emissions and adding to packaging waste.

4. **Use and Consumption**
   During this phase, consumers use the product. The environmental impact of this stage depends on how the product is used. For example, the energy consumed by appliances, the water used for personal care products, or the carbon emissions from a vehicle all contribute to the product's overall impact.

   Example: A plastic bottle might only be used for a few minutes, but the environmental cost of its production, transportation, and disposal is far greater. Similarly, electric appliances that use large amounts of energy contribute to higher emissions if the energy is derived from fossil fuels.

5. **Disposal and Waste Management**
   The final stage of the life cycle is disposal, where waste is either recycled, incinerated, or sent to landfills. How waste is managed affects its environmental impact. Landfills emit harmful methane gases, incineration releases toxins into the air, and recycling can save energy and reduce the need for new raw materials. However, recycling is not always as efficient or widely practiced as it should be.

   Example: Plastic that ends up in a landfill can take hundreds of years to break down, during which time it leaches chemicals into the soil and water. On the other hand, recycling paper and glass saves significant amounts of energy compared to producing them from raw materials.

The Environmental Impact of Waste

The environmental impact of waste is vast and often hidden from view. Understanding these impacts helps emphasize the importance of reducing waste at every stage of its life cycle.

1. **Pollution**
   Waste—especially plastic, chemicals, and non-biodegradable materials—can pollute land, water, and air. When waste is improperly disposed of, it can leach harmful chemicals into the environment. Plastics, in particular, contribute to marine pollution, with millions of tons of plastic ending up in the oceans every year, harming marine life and ecosystems.

   Example: Microplastics, which are tiny plastic particles, can enter the food chain through marine life, affecting everything from plankton to fish and even humans.

2. **Greenhouse Gas Emissions**
   Landfills and incinerators are major sources of greenhouse gas emissions. As organic materials decompose in landfills, they release methane, a potent greenhouse gas. Similarly, burning waste in incinerators releases carbon dioxide, contributing to global warming. The energy required for waste management—transportation, processing, and disposal—also adds to the carbon footprint.

   Example: A single ton of municipal waste sent to a landfill can produce about 1 ton of CO_2 equivalent in emissions. Burning waste in an incinerator, depending on the technology used, can produce additional harmful gases and contribute to air pollution.

3. **Resource Depletion**
   The production of goods requires raw materials, many of which are finite and non-renewable. When products are discarded instead of reused or recycled, these resources are lost, and new raw materials must be extracted, contributing to environmental degradation and resource depletion.

   Example: The mining of metals for electronics such as phones and computers leads to depletion of valuable minerals, which takes hundreds of years to regenerate.

4. **Loss of Biodiversity**
   Waste can directly or indirectly lead to the loss of biodiversity. Pollution from waste can contaminate habitats, poison species, and disrupt ecosystems. Additionally, resource extraction—necessary to produce new products—often results in deforestation and destruction of wildlife habitats.

Example: Illegal dumping of toxic waste near rivers can destroy aquatic ecosystems, harming fish and plant life. Deforestation caused by mining and agriculture to create products contributes to the endangerment of species that depend on these habitats.

The Importance of Waste Reduction

The environmental impact of waste is undeniable, which is why reducing waste is one of the most effective ways to minimize our footprint on the planet. By focusing on waste reduction, we can lower greenhouse gas emissions, conserve resources, reduce pollution, and protect biodiversity.

Key Takeaways

- **Waste is a life cycle**: From raw material extraction to disposal, every stage of a product's life contributes to its environmental impact.
- **The environmental cost is high**: Waste contributes to pollution, greenhouse gas emissions, resource depletion, and loss of biodiversity.
- **Reducing waste matters**: Minimizing waste is crucial for conserving resources, reducing pollution, and slowing climate change.

By understanding the life cycle of waste and its environmental impact, we can begin to make more conscious decisions in our daily lives. The next section will introduce the core strategies for reducing waste, including the 5Rs: Refuse, Reduce, Reuse, Recycle, and Rot. These simple but powerful strategies can guide you in making more sustainable choices and reducing your environmental footprint.

2. The 5Rs: Refuse, Reduce, Reuse, Recycle, Rot

Reducing waste is one of the most effective ways to live sustainably and protect the planet. The 5Rs—Refuse, Reduce, Reuse, Recycle, and Rot—are core strategies that can help guide you in minimizing waste at every stage of its life cycle. By following these principles, you can make conscious choices to lessen your environmental impact and contribute to a healthier planet.

Each of the 5Rs focuses on a specific aspect of waste reduction, and when combined, they form a powerful framework for sustainable living. Let's take a closer look at each of these strategies:

1. Refuse: Say No to Waste

The first and most powerful step in waste reduction is to **refuse** items and practices that contribute to unnecessary waste. Refusing is about being mindful of what you consume and avoiding things that you don't need. This can mean saying no to single-use plastics, excessive packaging, and disposable products.

Examples of Refusing:

- **Refuse plastic straws**: Bring your own reusable straw or simply avoid using straws.
- **Refuse unnecessary packaging**: When shopping, choose products with minimal or recyclable packaging.
- **Refuse freebies**: Many companies give out promotional items like keychains, plastic bags, or flyers. Instead of accepting these items, politely refuse them.
- **Refuse fast fashion**: Avoid buying cheap, low-quality clothes that contribute to landfill waste by opting for sustainable fashion choices.
- **Refuse disposable coffee cups**: Bring your own reusable coffee cup when you go to your local café to avoid the waste generated by disposable cups.

By refusing, you stop waste before it even starts. This step requires being mindful and intentional about your choices. Refusing to buy or accept products with excessive waste helps to reduce the demand for these items and curtails their environmental impact.

2. Reduce: Minimize Your Consumption

Once you've begun refusing unnecessary waste, the next step is to **reduce** the amount of waste you generate in the first place. This means cutting back on consumption, avoiding overbuying, and choosing products that have a longer life span or can be used in multiple ways.

Examples of Reducing:

- **Reduce energy usage**: Turn off lights when you leave a room, unplug devices when they're not in use, and use energy-efficient appliances.
- **Reduce water usage**: Take shorter showers, fix leaks, and use water-efficient appliances like low-flow showerheads or dishwashers.
- **Reduce food waste**: Buy only what you need, plan meals to avoid leftovers, and use up ingredients before they go bad.
- **Reduce car usage**: Walk, bike, or use public transportation instead of driving short distances.

- **Reduce the use of paper**: Opt for digital documents and e-bills instead of printed copies when possible.

By reducing your consumption, you cut down on the amount of waste you generate and reduce your overall environmental impact. This not only helps the planet but can also save you money and resources in the long run.

3. Reuse: Give Items a Second Life

The next step is **reuse**—making sure that the things you already have continue to serve a useful purpose instead of being discarded. Reusing extends the life of products and reduces the need for new resources to make replacements.

Examples of Reusing:

- **Reuse glass jars and containers**: Instead of throwing away glass jars from jam or sauces, wash them and repurpose them for storage, organization, or even as drinking glasses.
- **Reuse shopping bags**: Keep reusable bags in your car, purse, or backpack so you can use them for shopping instead of accepting plastic bags at stores.
- **Reuse clothing**: Buy second-hand clothes, repair damaged garments, and donate or sell items you no longer wear rather than throwing them away.
- **Reuse packaging**: If you receive products in boxes or bubble wrap, save them for future use when shipping or storing items.
- **Repurpose old furniture**: Give your furniture a makeover by sanding, painting, or reupholstering instead of discarding it.

By reusing, you get more value from the things you already own and help reduce the demand for new products. Reuse also helps reduce the energy, resources, and waste associated with manufacturing, packaging, and transportation.

4. Recycle: Turn Waste into New Resources

Recycling involves converting waste materials into new products or raw materials, preventing valuable resources from ending up in landfills. This is an important step in the waste management process and can help reduce the environmental impact of extraction and manufacturing.

Examples of Recycling:

- **Recycling paper and cardboard**: These materials can be turned into new paper products, reducing the need to cut down trees.
- **Recycling plastics**: Many types of plastic can be recycled into new plastic products, reducing the amount of new plastic that needs to be produced.
- **Recycling metals**: Metals like aluminum and steel can be endlessly recycled, saving energy and resources compared to extracting and processing new metal ores.
- **Recycling electronics**: Electronics contain valuable metals like gold, silver, and copper. Many old electronics can be recycled for parts and raw materials.
- **Recycling glass**: Glass can be recycled infinitely without losing quality, making it a great candidate for the recycling process.

To recycle effectively, make sure you follow your local recycling guidelines and sort materials appropriately. While recycling is an important strategy, it should be viewed as a last resort—once you've taken steps to refuse, reduce, and reuse.

5. Rot: Compost Organic Waste

The final step in the 5Rs is **rot**, which involves composting organic waste instead of sending it to a landfill. Composting is the process of breaking down organic materials, like food scraps and yard waste, into nutrient-rich soil that can be used to enrich gardens or farms.

Examples of Rotting (Composting):

- **Composting food scraps**: Vegetable peels, coffee grounds, and fruit cores can be composted to create rich soil that can be used for gardening.
- **Composting yard waste**: Grass clippings, leaves, and branches can be composted, reducing the need for chemical fertilizers.
- **Composting paper**: Paper towels, napkins, and paper plates (without plastic coatings) can be composted to further reduce waste.

By composting organic materials, you divert food scraps and yard waste from landfills, where they would otherwise produce methane (a potent greenhouse gas). Composting also helps reduce the need for synthetic fertilizers and encourages healthier soil for gardening and farming.

Key Takeaways

- **Refuse** to accept unnecessary items or products that create waste.
- **Reduce** your consumption by choosing fewer, higher-quality items and using them for longer periods.
- **Reuse** items by repurposing them for other purposes instead of throwing them away.
- **Recycle** materials that cannot be reused, ensuring they are processed into new products or raw materials.
- **Rot** by composting organic waste, turning it into valuable resources for gardening or farming.

By following the 5Rs—Refuse, Reduce, Reuse, Recycle, and Rot—you can make a significant impact on the amount of waste you generate and reduce your environmental footprint. In the next section, we'll explore practical ways to reduce household waste, including tips for waste reduction in the kitchen, bathroom, and daily life.

3. Practical Ways to Reduce Household Waste

Reducing waste at home is one of the most effective ways to live sustainably. By making simple changes in everyday habits, you can significantly reduce the amount of waste your household generates. The key to waste reduction is being mindful of what you consume, how you dispose of it, and how you reuse or repurpose items. In this section, we'll look at practical tips for minimizing waste in key areas of your home—kitchen, bathroom, and daily life.

In the Kitchen

The kitchen is one of the most waste-intensive areas of the home, with food packaging, food waste, and disposable products contributing to the problem. However, there are numerous ways to cut down on kitchen waste and make more sustainable choices.

1. Reduce Food Waste
Food waste is a major contributor to landfill waste and greenhouse gas emissions. By being mindful of food purchasing, storage, and preparation, you can drastically reduce food waste.

- **Plan meals and shop with a list**: By planning meals ahead of time and sticking to a shopping list, you can avoid buying excessive amounts of food that might go unused.
- **Store food properly**: Learn how to store fruits, vegetables, dairy, and grains to extend their shelf life. For example, store fruits like apples and bananas in the fridge to prevent them from ripening too quickly,

and keep leafy greens in breathable bags or containers to maintain freshness.
- **Use leftovers creatively**: Repurpose leftover meals into new dishes. For example, roast vegetables can be turned into a soup or used as a salad topping. Leftover grains can be added to stir-fries or salads.
- **Compost food scraps**: Instead of throwing away food scraps like vegetable peels, coffee grounds, or eggshells, compost them to create nutrient-rich soil for your garden.

2. Choose Minimal Packaging

Packaging, especially single-use plastic, is a significant source of waste in the kitchen. Reducing packaging can help minimize waste.

- **Buy in bulk**: Purchase dry goods like rice, pasta, and grains from bulk bins to avoid excess packaging. Bring your own reusable containers or bags to fill up.
- **Avoid pre-cut or pre-wrapped produce**: Instead of buying pre-packaged vegetables or fruits, buy whole produce to reduce packaging waste.
- **Use reusable produce bags**: When buying fruits and vegetables, use cloth or mesh bags instead of the plastic bags provided by the store.
- **Opt for glass or metal containers**: Choose glass jars, metal tins, or compostable materials for storing food at home instead of single-use plastic containers.

3. Reduce Single-Use Items

The kitchen is home to many single-use items that contribute to waste. Consider switching to reusable alternatives.

- **Use reusable shopping bags**: Say goodbye to plastic bags by using reusable fabric bags when you go grocery shopping.
- **Switch to cloth napkins**: Instead of using paper napkins or paper towels, use cloth napkins that can be washed and reused.
- **Invest in reusable food wraps**: Replace plastic wrap and aluminum foil with reusable beeswax wraps or silicone lids.
- **Switch to a refillable water bottle**: Avoid single-use plastic bottles by using a durable, refillable bottle.

In the Bathroom

The bathroom is another area of the home that generates a lot of waste, mainly due to disposable personal care products and packaging. Here are some practical tips to reduce waste in your bathroom.

1. Choose Reusable Personal Care Products

Many bathroom products are designed to be disposable, but there are plenty of reusable alternatives that can help you reduce waste.

- **Switch to a reusable razor**: Instead of disposable razors, invest in a high-quality reusable razor with replaceable blades. This will drastically cut down on razor waste.
- **Use cloth or bamboo cotton pads**: Swap out disposable cotton pads for reusable cloth pads or bamboo pads for makeup removal and skincare routines.
- **Opt for a menstrual cup or reusable pads**: Traditional menstrual products like pads and tampons create significant waste. Consider switching to a reusable menstrual cup or cloth pads, which can be washed and reused.
- **Buy bar soap instead of liquid soap**: Bar soap often has less packaging than liquid soap and lasts longer, making it a more eco-friendly option.

2. Reduce Plastic and Excess Packaging

Packaging waste in the bathroom is primarily from single-use plastic bottles, tubes, and containers. By opting for products with minimal or recyclable packaging, you can help reduce this waste.

- **Choose products with minimal packaging**: Look for brands that offer products with little or no plastic packaging, such as bar soap, shampoo bars, and natural deodorants in cardboard or metal containers.
- **Refill products**: Many stores offer refill stations for products like hand soap, shampoo, and body wash. Consider buying refillable containers and refilling them at these stations to reduce packaging waste.
- **Switch to toothpaste tablets or powder**: Instead of traditional toothpaste tubes, consider using toothpaste tablets or powder, which typically come in recyclable or compostable packaging.

3. Avoid Single-Use Bathroom Items

Many bathroom products are designed for single use, but by choosing reusable or longer-lasting alternatives, you can significantly reduce waste.

- **Use reusable toilet paper alternatives**: If you're feeling adventurous, consider trying reusable toilet paper options, like family cloths, which can be washed and reused.
- **Choose bamboo or compostable toothbrushes**: Instead of plastic toothbrushes, opt for bamboo toothbrushes, which are biodegradable.

- **Opt for natural loofahs**: Replace synthetic sponges with natural loofahs, which are compostable when they wear out.

In Daily Life

Beyond the kitchen and bathroom, there are many everyday habits and practices that can contribute to reducing household waste. Here are some tips for minimizing waste in daily life.

1. Reduce Paper Waste
Paper products, such as bills, receipts, and notebooks, often contribute to unnecessary waste.

- **Go paperless**: Switch to digital billing and statements to reduce the need for paper. Opt for email receipts instead of printed ones when shopping.
- **Reuse scrap paper**: Before tossing scrap paper, reuse it for notes or as a notepad for quick reminders.
- **Switch to cloth towels and rags**: Replace paper towels and napkins with cloth alternatives that can be washed and reused.

2. Be Mindful of Fashion and Clothing
Clothing production and disposal contribute to massive amounts of waste. By making conscious decisions about what you wear and how you care for your clothes, you can reduce your clothing waste.

- **Buy second-hand clothes**: Shopping at thrift stores or online second-hand marketplaces reduces the demand for new clothing and minimizes textile waste.
- **Invest in quality, long-lasting pieces**: Instead of following trends and buying low-quality fast fashion, invest in durable, timeless pieces that will last longer.
- **Repair or repurpose old clothing**: Instead of throwing out clothes that are worn out or damaged, consider repairing them or repurposing them into something new, like turning old t-shirts into cleaning rags or bags.

3. Reduce Electronic Waste
Electronic devices and gadgets are often discarded when they break or become outdated. However, there are ways to extend the life of your electronics and minimize electronic waste.

- **Repair instead of replace**: If your phone, laptop, or appliance breaks, try to repair it before buying a new one. Many electronics can be fixed or refurbished to extend their lifespan.
- **Recycle electronics responsibly**: When it's time to replace your electronics, make sure to recycle them properly by finding a certified e-waste recycling facility.
- **Avoid disposable batteries**: Use rechargeable batteries for devices like remotes, flashlights, and toys to reduce the need for single-use batteries.

Key Takeaways

- **Plan and shop smart** to reduce food waste and minimize packaging.
- **Use reusable alternatives** in the kitchen and bathroom, such as cloth towels, glass containers, and refillable products.
- **Choose products with minimal packaging** or buy in bulk to reduce excess waste.
- **Embrace repair and reuse** in all aspects of daily life, from clothing to electronics.
- **Compost food scraps** and other organic waste to divert it from landfills and enrich your garden.

By implementing these practical tips in your kitchen, bathroom, and daily routines, you can significantly reduce the waste your household produces. Small changes add up, and with time, you'll develop habits that align with a more sustainable, eco-friendly lifestyle.

4. Switching to Reusables: Alternatives to Single-Use Products

One of the most effective ways to reduce waste and live sustainably is to **switch to reusable alternatives** for items that are typically single-use. Single-use products, like plastic bags, straws, and disposable coffee cups, contribute significantly to pollution and landfill waste. Fortunately, there are countless reusable options available that can replace these disposable items in your daily life. In this section, we'll explore practical alternatives to commonly used single-use products and how they can help you make a lasting environmental impact.

1. Reusable Bags Instead of Plastic Bags

Plastic bags are one of the most common single-use items that contribute to waste. They take hundreds of years to decompose and often end up in the ocean, harming wildlife. By switching to reusable bags, you can make a significant difference in reducing plastic waste.

- **Cloth or Jute Bags**: Bring your own cloth, jute, or canvas bags to the store when shopping. These are sturdy and can hold more items than plastic bags.
- **Foldable Reusable Bags**: Invest in foldable reusable bags that you can easily carry in your purse or pocket. This ensures that you always have a bag when you need it.
- **Avoid Produce Bags**: Many reusable grocery bags come with compartments for produce, eliminating the need for disposable plastic produce bags. If not, you can use reusable mesh or cotton produce bags.

2. Reusable Water Bottles Instead of Single-Use Bottled Water

Single-use plastic water bottles are a major source of plastic pollution, as they are often used for just a few minutes and then discarded. Switching to a reusable water bottle is one of the easiest and most impactful changes you can make.

- **Stainless Steel or Glass Bottles**: Opt for a high-quality stainless steel or glass water bottle, which will last for years and keep your water cool or hot for longer periods.
- **Filtered Bottles**: Some reusable water bottles come with built-in filters, allowing you to drink clean water from almost any source, reducing the need for bottled water.
- **Insulated Bottles**: Insulated bottles help maintain the temperature of your drink, making them ideal for both hot beverages like coffee and cold drinks like iced tea.

3. Reusable Coffee Cups Instead of Disposable Coffee Cups

Disposable coffee cups, lids, and stirrers are used for only a short time but contribute to enormous amounts of waste. Switching to a reusable coffee cup is a simple yet effective step in reducing waste.

- **Travel Mugs**: Stainless steel, bamboo, or insulated travel mugs are perfect for coffee or tea on the go. Many coffee shops even offer discounts to customers who bring their own cups.
- **Reusable Lids and Cups**: Look for reusable coffee cups made from materials like glass, silicone, or stainless steel. These are easy to clean and can last for years.

- **Reusable Coffee Filters**: If you brew coffee at home, consider using a reusable coffee filter instead of disposable paper ones. These are easy to wash and can be used for hundreds of brews.

4. Reusable Straws Instead of Plastic Straws

Plastic straws are another major contributor to plastic pollution, especially in marine environments. Many people don't realize that straws can be easily replaced with reusable alternatives.

- **Metal Straws**: Stainless steel straws are durable, easy to clean, and come in various sizes to fit different cups and bottles.
- **Bamboo Straws**: Bamboo straws are a biodegradable and sustainable option, perfect for anyone looking for a more eco-friendly alternative.
- **Silicone Straws**: Silicone straws are soft, flexible, and dishwasher-safe, making them a great option for families with young children or for those who prefer a more comfortable drinking experience.
- **Glass Straws**: Glass straws are a stylish and reusable alternative. While they require careful handling, they are easy to clean and are free of chemicals often found in plastic.

5. Reusable Food Storage Instead of Plastic Wrap and Ziploc Bags

Plastic wraps, sandwich bags, and aluminum foil are commonly used for storing food but are often single-use and harmful to the environment. Reusable alternatives help reduce waste and keep your food fresh.

- **Beeswax Wraps**: Beeswax wraps are a natural and compostable alternative to plastic wrap. They are perfect for wrapping fruits, vegetables, and sandwiches and can be reused multiple times.
- **Silicone Food Storage Bags**: Reusable silicone bags can be used for storing snacks, sandwiches, or leftovers. They are durable, easy to clean, and come in various sizes to meet your storage needs.
- **Glass or Stainless Steel Containers**: Replace single-use plastic food containers with reusable glass or stainless steel containers. These options are great for meal prep and are microwave- and dishwasher-safe.
- **Fabric Bowl Covers**: Instead of using plastic wrap to cover bowls, use cloth bowl covers that can be washed and reused.

6. Reusable Napkins Instead of Paper Napkins

Disposable paper napkins are used once and then thrown away, but cloth napkins can be washed and reused many times. Switching to reusable napkins is an easy and stylish way to reduce paper waste.

- **Cloth Napkins**: Invest in a set of durable cloth napkins made from cotton, linen, or bamboo fabric. These are easy to wash and come in a variety of colors and patterns to match your home.
- **Reusable Paper Towels**: Cloth napkins or reusable "paper towels" made from fabric can be used to clean up messes and spills in the kitchen. Simply toss them in the wash after use.

7. Reusable Shopping and Produce Bags Instead of Plastic Bags

Plastic bags are commonly used for carrying items from the store, but they contribute significantly to waste. Reusable bags are a sustainable alternative that can help cut down on plastic consumption.

- **Mesh Produce Bags**: For fruits and vegetables, use reusable mesh or cotton produce bags instead of the plastic ones provided by stores.
- **Multi-Purpose Reusable Bags**: Invest in high-quality, durable reusable bags that can carry groceries, books, and even clothes. Many of these bags fold up compactly so you can carry them wherever you go.

8. Reusable Diapers Instead of Disposable Diapers

Disposable diapers are a major source of waste, with millions of diapers ending up in landfills each day. Though it requires a bit more effort, switching to reusable cloth diapers can make a huge difference.

- **Cloth Diapers**: Modern cloth diapers are more convenient and effective than ever. They come in a variety of sizes and designs and can be washed and reused multiple times, reducing waste.
- **Diaper Liners**: Use washable cloth liners inside cloth diapers to make cleanup easier. These liners can be removed and washed separately, reducing the need for disposable diaper products.

9. Reusable Cleaning Cloths Instead of Paper Towels

Paper towels are commonly used for cleaning but generate a lot of waste. By switching to reusable cleaning cloths, you can reduce the amount of paper waste generated from cleaning tasks.

- **Microfiber Cloths**: Microfiber cloths are highly absorbent and effective at cleaning surfaces without the need for harsh chemicals.
- **Cotton Rags**: Repurpose old clothes, towels, or sheets as cleaning rags to avoid buying new paper towels. These rags can be used over and over again.

Key Takeaways

- **Reusable bags** help reduce plastic bag waste, whether for groceries, produce, or general shopping.
- **Reusable water bottles** reduce reliance on single-use plastic bottles, saving money and the environment.
- **Reusable coffee cups and straws** cut down on the disposable coffee cup waste generated by cafes and take-out beverages.
- **Reusable food wraps and containers** help eliminate the need for plastic wraps and sandwich bags.
- **Cloth napkins and towels** can replace paper napkins and paper towels in the kitchen and bathroom.

By making these simple switches to reusable alternatives, you can significantly reduce your environmental footprint. Each time you use a reusable product, you help minimize the waste that ends up in landfills and the pollution that harms the planet. Make the change today, and you'll be one step closer to a more sustainable lifestyle.

Chapter 3: Sustainable Food Choices

1. Why Food Choices Matter

The food we consume has a significant impact on the environment, far beyond what most people realize. From the resources used to grow food to how it's transported and disposed of, every step of the food production process leaves a footprint. In this section, we'll explore the environmental consequences of our food choices and why making sustainable food decisions is critical to protecting the planet.

The Environmental Impact of Our Diet

Our daily food choices are connected to several environmental issues, including climate change, deforestation, water use, and biodiversity loss. While food is essential for our survival, the way it is produced, processed, packaged, and consumed can either harm or help the planet. Understanding the environmental impact of our food choices can empower us to make decisions that align with our values and contribute to a more sustainable future.

1.1 Greenhouse Gas Emissions and Climate Change

The global food system is responsible for a significant portion of greenhouse gas (GHG) emissions, which contribute to climate change. Livestock farming, in particular, has a large carbon footprint due to the methane produced by animals, the energy-intensive processes of feed production, and the large-scale deforestation required to create grazing lands and grow animal feed.

- **Animal agriculture**: Meat and dairy production, especially beef and lamb, have particularly high GHG emissions because ruminant animals (such as cows and sheep) release methane during digestion, a potent greenhouse gas. In addition, raising animals for food requires large amounts of land, water, and feed.
- **Plant-based foods**: On the other hand, plant-based foods tend to have a much lower carbon footprint. For example, growing vegetables, grains, legumes, and fruits generally requires fewer resources and results in far fewer emissions compared to animal products.

Takeaway: Choosing plant-based foods, such as vegetables, grains, legumes, and nuts, can significantly lower your carbon footprint. Reducing your consumption of high-emission foods like red meat and dairy can make a big difference in combating climate change.

1.2 Land Use and Deforestation

The food industry is a major driver of deforestation and habitat destruction. As the demand for food, especially animal products, increases, forests are cleared to make way for farming. This land-use change not only results in the loss of valuable ecosystems but also reduces the planet's ability to absorb carbon dioxide.

- **Livestock farming**: Cattle farming, in particular, is a leading cause of deforestation in areas such as the Amazon rainforest. Forests are cleared to create pastureland for grazing or to grow feed crops like soy.
- **Crop production**: While crop production doesn't directly contribute to deforestation in the same way as livestock farming, large-scale monocropping (such as growing corn or soy) can lead to soil depletion, water pollution, and the loss of biodiversity.

Takeaway: Supporting sustainable farming practices and choosing foods that are grown using agroecological methods can help protect natural ecosystems. Opting for plant-based foods and choosing products certified as sustainable (e.g., organic, Fair Trade) can help mitigate deforestation.

1.3 Water Use and Scarcity

Water is an essential resource in food production, but the way we grow our food has a significant impact on global water supplies. Some food products are incredibly water-intensive to produce, while others require far less.

- **Water-intensive foods**: Animal agriculture is extremely water-hungry, especially in the case of beef. It takes thousands of liters of water to produce a single kilogram of beef, as water is needed not only for the animals themselves but also for growing the feed.
- **Plant-based foods**: Growing plants generally requires less water than raising livestock. Crops like grains, beans, and vegetables have a much lower water footprint.

Takeaway: Reducing your consumption of water-intensive animal products, particularly beef, and shifting to plant-based foods can help conserve precious water resources. Additionally, choosing locally grown and seasonal foods can reduce the amount of water used for irrigation, as these foods are better adapted to the local climate.

1.4 Biodiversity and Soil Health

The food we eat also affects biodiversity—the variety of plant and animal life that makes our ecosystems thrive. Monocropping, pesticide use, and the expansion of agriculture into wild habitats can harm biodiversity and soil health, which in turn affects food security and ecosystem stability.

- **Monocropping**: The practice of growing a single crop over vast areas reduces soil fertility and makes crops more vulnerable to pests and diseases. It also limits the variety of plants and animals in the ecosystem.
- **Pesticides and chemicals**: The widespread use of chemical fertilizers and pesticides in conventional agriculture harms beneficial insects (like bees), disrupts natural ecosystems, and pollutes waterways.
- **Agroecology and regenerative farming**: Sustainable farming methods like agroecology and regenerative agriculture prioritize soil health, biodiversity, and ecosystem restoration. These methods use crop rotation, composting, and integrated pest management to protect the environment.

Takeaway: By choosing organic and sustainably produced food, you help support farming practices that maintain biodiversity, promote soil health, and protect natural ecosystems. Avoiding highly processed foods and products from industrial farming systems is a key step in preserving the planet's biodiversity.

1.5 Waste and Food Loss

Another environmental issue tied to food choices is food waste. A large portion of food produced globally is wasted—either during production, transportation, retail, or at the consumer level. Wasting food means wasting the resources (water, energy, labor) that went into producing it.

- **Food waste in landfills**: When food waste ends up in landfills, it decomposes anaerobically, producing methane gas, which contributes to climate change. Additionally, the resources used to grow, harvest, and transport that food are also wasted.
- **Reducing food waste**: By reducing food waste through smarter shopping, meal planning, proper storage, and using leftovers creatively, we can reduce the environmental impact of food production and help alleviate food insecurity.

Takeaway: Reducing food waste is one of the most impactful actions you can take to lower your environmental footprint. Be mindful of portion sizes, store food properly, and use leftovers to minimize waste.

Key Takeaways

- **Animal agriculture has a high environmental impact**, especially in terms of greenhouse gas emissions, deforestation, and water usage. Reducing meat and dairy consumption can have a significant positive impact on the environment.
- **Plant-based foods have a lower environmental footprint** in terms of greenhouse gas emissions, land use, and water consumption.
- **Sustainable farming practices** that prioritize soil health, biodiversity, and reduced pesticide use help protect ecosystems and promote long-term food security.
- **Reducing food waste** by planning meals, storing food properly, and using leftovers can help conserve the resources used in food production.

By understanding the environmental impact of our food choices, we can make informed decisions that not only benefit our health but also contribute to the well-being of the planet. Every meal is an opportunity to align our diet with our environmental values, and small changes can add up to make a significant difference.

2. Shopping Mindfully

In today's world, food packaging contributes significantly to waste, especially single-use plastic. However, with a few simple changes in how we shop, we can significantly reduce packaging waste and make more sustainable choices. In this section, we'll explore ways to shop mindfully by reducing packaging, buying in bulk, and making eco-conscious decisions at the store.

2.1 Bring Your Own Containers

One of the most effective ways to reduce packaging waste is by **bringing your own containers** when you go shopping. Many stores now offer the option to bring your own jars, bags, and containers to buy bulk items, such as grains, nuts, dried fruits, and spices.

- **Reusable Glass Jars and Containers**: Invest in a set of glass jars or containers for storing bulk items. Glass is durable, reusable, and doesn't leach chemicals into food like some plastics can. You can use these containers for everything from dried beans to bulk grains.
- **Cloth Bags for Produce**: Instead of using plastic bags for produce, opt for reusable cloth or mesh produce bags. These bags are lightweight and breathable, making them perfect for fruits, vegetables, and herbs.

- **Zero-Waste Shops**: Many cities now have zero-waste stores where you can fill your containers directly from bulk bins. These shops typically carry products like rice, pasta, cereals, and even cleaning supplies in bulk, without the excess packaging.

Tip: When shopping in bulk, make sure to label your containers with the weight before filling them, and keep track of your store's tare weight policy (the weight of the container itself).

2.2 Buy in Bulk

Buying in bulk is one of the easiest and most impactful ways to **reduce packaging waste**. Bulk buying means purchasing larger quantities of goods, typically from bulk bins or dispensers, which reduces the need for individually packaged items. This method also saves you money in the long run.

- **Grains, Legumes, and Pasta**: Look for bulk bins in your local grocery store or bulk food stores for staples like rice, oats, lentils, beans, and pasta. You'll not only reduce packaging but also save money compared to buying pre-packaged items.
- **Snacks and Nuts**: Many stores now offer nuts, seeds, dried fruits, and snacks in bulk, often at a better price than their packaged counterparts. Use your reusable containers to fill up with exactly the amount you need.
- **Cleaning Products and Toiletries**: Many sustainable brands offer bulk refills for cleaning products, soaps, shampoos, and conditioners. These refills often come in large containers that you can reuse, which is much more eco-friendly than buying new bottles every time.
- **Frozen Food**: If possible, buy frozen fruits and vegetables in bulk to avoid individually packaged items. Larger, bulk bags are usually more sustainable and economical.

Tip: Start small by purchasing products that are commonly available in bulk, like rice, flour, and nuts. Gradually expand to other categories as you become more accustomed to shopping this way.

2.3 Avoid Single-Use Plastic Packaging

Single-use plastics, such as plastic wrap, plastic bags, and pre-packaged foods, contribute significantly to environmental pollution. When shopping, look for products with minimal or eco-friendly packaging, or better yet, avoid them altogether.

- **Choose Loose Produce**: Avoid buying fruits and vegetables that are wrapped in plastic. Instead, choose loose produce and use your reusable produce bags to carry them. This can save a huge amount of plastic waste over time.
- **Glass, Paper, or Cardboard Packaging**: Opt for products that come in glass, paper, or cardboard packaging, as these materials are more easily recyclable and biodegradable compared to plastic. For example, many brands now offer beverages, sauces, and condiments in glass bottles or jars.
- **Skip Individually Wrapped Snacks**: Instead of buying individually wrapped snacks or portions, opt for larger containers or bulk options that you can divide yourself at home into reusable bags or containers.
- **Avoid Pre-Packaged Meat**: Many stores offer fresh meat and fish that are pre-packaged in plastic. Look for butcher counters or the meat section where products are sold without plastic wrapping or ask if they can wrap your purchase in paper or place it directly into your reusable container.

Tip: Bring your own reusable bags for non-produce items, and opt for products with eco-friendly or recyclable packaging. Over time, this reduces your dependence on plastic bags and wraps.

2.4 Support Local and Sustainable Brands

Supporting local farmers, sustainable food brands, and eco-conscious businesses is another important aspect of mindful shopping. By purchasing from local and ethical sources, you're reducing the environmental impact of transportation and supporting practices that align with your values.

- **Farmers' Markets**: Shopping at farmers' markets is a great way to buy fresh produce, meats, and dairy directly from local farmers. Not only does this reduce the carbon footprint associated with transportation, but it also gives you the opportunity to ask questions about the farming practices used, such as whether they are organic or regenerative.
- **Community-Supported Agriculture (CSA)**: Many regions offer CSA programs where you can subscribe to receive seasonal produce directly from local farms. This supports local agriculture, reduces packaging waste, and ensures that you are eating in harmony with the seasons.
- **Sustainable Brands**: Look for brands that prioritize sustainability, such as those that use recycled packaging, adopt fair trade practices, or are certified organic. Many stores now carry sustainably sourced

products, such as coffee, tea, and chocolate, which have a much smaller environmental impact than conventionally farmed goods.

Tip: When buying packaged goods, always look for certification labels that indicate eco-friendly practices, such as Fair Trade, organic, or carbon-neutral certifications.

2.5 Plan Your Shopping List

Shopping mindfully is also about making intentional decisions before you even enter the store. By planning your meals and your shopping list in advance, you can avoid impulse purchases, reduce food waste, and minimize packaging.

- **Meal Planning**: Plan your meals for the week and create a shopping list based on the ingredients you need. This helps you avoid buying unnecessary items that could end up in the trash. It also prevents buying products with excessive packaging or items you won't use.
- **Inventory Check**: Before you go shopping, check your pantry and fridge to see what you already have. This reduces the chances of buying duplicate items, cutting down on both food waste and unnecessary packaging waste.
- **Bulk Shopping**: By knowing exactly what you need, you can purchase bulk items in the correct quantities, reducing overbuying and the packaging waste associated with smaller, pre-packaged portions.

Tip: Having a meal plan and a shopping list also helps you avoid fast food or takeout, which often comes in excessive packaging, and ensures that your meals are more sustainable and less reliant on pre-packaged products.

2.6 Reduce Impulse Purchases

Impulse buying is a common habit that can lead to excess packaging waste and unneeded products. Mindful shopping is all about being intentional with your purchases and focusing on what you truly need.

- **Stick to Your List**: A well-prepared shopping list helps you avoid purchasing items on a whim. Keep your list simple and focused on necessary, sustainable products.
- **Avoid Aisles with Excess Packaging**: When shopping, try to avoid the aisles that feature overly packaged, processed, or convenience foods. Instead, stick to the produce section and bulk bins to find fresh, unprocessed food.

- **Mindful Choices**: If you do see something that isn't on your list but catches your eye, ask yourself whether it's something you truly need or whether it will just add unnecessary packaging to your waste.

Key Takeaways

- **Bring your own containers** to shop in bulk and reduce packaging waste.
- **Buy in bulk** for staples like grains, nuts, and snacks to avoid individual packaging.
- **Avoid single-use plastic** by choosing products in glass, paper, or biodegradable packaging.
- **Support local and sustainable brands** to reduce your food's carbon footprint and packaging waste.
- **Plan your shopping list** to reduce waste, prevent impulse buying, and limit unnecessary packaging.
- **Stick to your list** to avoid buying items that will contribute to packaging waste.

By adopting mindful shopping habits, we can dramatically reduce the amount of waste we generate and make more sustainable choices. Small changes in the way we shop add up over time and can help make a lasting difference in the fight against environmental pollution.

3. Eating Seasonal and Local

Eating seasonally and locally is one of the most effective ways to reduce your environmental impact, support local economies, and enjoy fresh, nutrient-rich food. This approach not only aligns with eco-friendly living but also helps you connect more deeply with the cycles of nature and the food production process. In this section, we'll explore the benefits of eating seasonal and local, and how you can incorporate this practice into your everyday life.

3.1 What Does "Seasonal Eating" Mean?

Seasonal eating refers to choosing fruits, vegetables, and other foods that are naturally harvested at certain times of the year. These foods are often fresher, tastier, and more affordable than those grown out of season, and they have a lower environmental impact due to less need for artificial heating, pesticides, or long-distance transportation.

- **Seasonal Foods**: Each season has its own bounty of fresh produce. For example, in the spring, you might find asparagus, peas, and

strawberries, while in the fall, squashes, apples, and root vegetables are abundant.
- **Seasonal Eating and Nutrition**: Foods that are grown in their natural season often retain more nutrients because they ripen at their peak, unlike out-of-season produce that may be harvested prematurely for shipping.

Takeaway: Eating seasonal foods means you're enjoying produce at its freshest and most nutritious, while also supporting farming practices that work in harmony with nature's cycles.

3.2 The Environmental Benefits of Eating Seasonal and Local

Eating seasonally and locally reduces the environmental impact of your food choices in several ways, from cutting down on food miles to supporting sustainable farming practices.

- **Reduced Carbon Footprint**: When you buy foods that are grown locally, you reduce the need for long-distance transportation. The environmental cost of shipping produce from across the globe, especially out-of-season items, can be immense. Foods grown locally require less fuel for transportation, reducing greenhouse gas emissions associated with food distribution.
- **Less Energy Intensive**: Foods grown out of season often require artificial heating, pesticides, and fertilizers to grow, which leads to higher energy consumption and a larger environmental footprint. In contrast, seasonal foods are typically grown with fewer inputs and are better adapted to the local climate.
- **Support for Sustainable Farming**: Local farmers who focus on seasonal produce are often more committed to using sustainable and regenerative farming practices. These methods prioritize soil health, biodiversity, and minimal chemical use, making them more eco-friendly in the long run.

Takeaway: By eating seasonal and local foods, you help reduce the carbon emissions related to food transportation, minimize energy use in agriculture, and support sustainable farming practices that protect the planet.

3.3 Supporting Local Farmers and Communities

Buying locally grown food directly supports the farmers and communities in your region, helping to build a more resilient food system.

- **Economic Benefits**: When you buy from local farmers, a greater portion of your money stays within the community, helping to support local economies. This is particularly important in rural or agricultural areas where farming is the main industry.
- **Building Community Connections**: Shopping locally also helps build connections between consumers and farmers. Many local farmers sell directly to consumers through farmers' markets, farm stands, and community-supported agriculture (CSA) programs, which allows you to learn more about where your food comes from and how it is grown.
- **Food Security**: Supporting local farms can help improve food security in your community. By choosing to purchase from small-scale, sustainable farmers, you are helping to strengthen the local food system and reduce the reliance on industrial agriculture and its environmental costs.

Takeaway: Supporting local farmers ensures that your food dollars contribute to your community's economy, while also fostering stronger connections to the food system.

3.4 How to Eat Seasonally and Locally

Incorporating more seasonal and local foods into your diet is easier than you might think. Here are some practical tips for making it a regular part of your life:

- **Shop at Farmers' Markets**: Farmers' markets are a great place to find fresh, seasonal produce. Vendors at these markets usually sell products grown locally, and you can often talk to the farmers directly to learn more about their growing practices.
- **Join a CSA**: Community-supported agriculture (CSA) programs allow you to subscribe to a weekly or monthly delivery of fresh, seasonal produce. Many CSAs focus on organic or sustainable farming methods, and you'll receive a rotating selection of fruits and vegetables that are harvested at their peak. This is a great way to experience a variety of seasonal foods and support local farms.
- **Learn the Seasons of Your Region**: Familiarize yourself with which foods are in season in your area. Local agricultural calendars can provide a guide to when certain fruits and vegetables are at their best. Many seasonal foods also have a specific harvest window, so knowing when they're available helps you make the most of them.
- **Grow Your Own**: If you have the space and time, growing your own fruits and vegetables is a great way to eat seasonally and locally. Even small-scale gardening can yield a variety of seasonal produce, such as

herbs, tomatoes, or leafy greens, which are easy to grow in many climates.
- **Plan Your Meals Around Seasonal Foods**: Try to base your weekly meals on what's in season. For example, in the fall, you might make soups and stews with root vegetables, while in the summer, you could create fresh salads with tomatoes, cucumbers, and herbs. When you plan your meals with the seasons in mind, you'll enjoy fresher, more sustainable food.

Takeaway: Start by shopping at local farmers' markets, joining a CSA, and learning what's in season in your area. Small adjustments like these can make a big difference in the sustainability of your food choices.

3.5 The Nutritional Benefits of Seasonal Foods

Seasonal foods are not only better for the environment—they can also be better for your health. Eating a variety of seasonal fruits and vegetables means you'll get a wide range of nutrients, as different foods are available at different times of the year.

- **Higher Nutrient Content**: Foods that are grown in-season tend to have higher levels of vitamins and minerals because they are allowed to ripen naturally on the plant. They are also more likely to be fresh when they reach your table, meaning they haven't lost their nutrients through extended storage or long transportation times.
- **More Flavorful**: Seasonal produce is often more flavorful because it is harvested at its peak ripeness. Eating with the seasons means you can enjoy foods at their most delicious, as fruits and vegetables will have a fuller taste and texture.
- **Better for Your Health**: Eating a diverse range of seasonal foods means you get more antioxidants, vitamins, and minerals, which can help boost your immune system and improve overall health.

Takeaway: Seasonal foods offer a range of health benefits, including better taste and higher nutrient levels. Eating fresh, local produce is a great way to improve your health while reducing your environmental impact.

Key Takeaways

- **Eating seasonally** means choosing foods that are grown and harvested during their natural growing season, which is often more nutritious and flavorful.

- **Local food** is better for the environment because it reduces transportation emissions and supports sustainable farming practices.
- **Support your local community** by purchasing from farmers' markets, CSAs, or local co-ops, which helps strengthen local economies and builds food security.
- **Plan meals around the seasons** to enjoy fresh, local produce and reduce the environmental impact of out-of-season foods.

By making seasonal and local eating a priority, you can reduce your environmental footprint, improve your health, and support local farmers and communities. Embrace the rhythms of nature and enjoy the bounty each season offers!

4. Reducing Food Waste

Food waste is one of the most pressing issues in modern society, not just because of the resources it wastes, but also due to the environmental toll it takes. In this section, we'll explore practical strategies to reduce food waste in your home, including how to store food properly, repurpose leftovers, and make the most out of every ingredient.

4.1 Understanding Food Waste

Before diving into solutions, it's important to understand just how much food is wasted and why it happens. According to the Food and Agriculture Organization (FAO), about **one-third of all food produced globally** is wasted or lost. Much of this waste occurs at the consumer level, often due to improper storage, mismanagement, or not using leftovers creatively.

Key Areas of Food Waste:

- **Improper Storage**: Many foods spoil prematurely when not stored correctly, leading to waste.
- **Over-Purchasing**: Buying more food than necessary or letting fresh produce sit in the fridge until it goes bad.
- **Plate Waste**: Serving too much food or cooking more than needed, which ends up being discarded.
- **Pre-Packaged Portions**: Many pre-portioned, single-use items end up going uneaten when not stored correctly or when they are not finished.

4.2 How to Store Food Properly

Proper food storage can significantly reduce spoilage, keeping your food fresh for longer and preventing unnecessary waste. Here are some tips for storing different types of food:

- **Fruits and Vegetables**:
    - **Refrigerate appropriately**: Not all fruits and vegetables need to be stored in the fridge. For example, apples, berries, and leafy greens should be refrigerated, while tomatoes, bananas, and potatoes do better in a cool, dark place.
    - **Use airtight containers**: Store chopped or cut vegetables and fruits in airtight containers to extend their shelf life.
    - **Keep ethylene-producing fruits separate**: Certain fruits, like apples, bananas, and avocados, produce ethylene gas, which can cause nearby fruits and vegetables to ripen and spoil faster. Keep them separate to prevent unnecessary waste.
- **Bread**:
    - **Freeze what you don't need immediately**: If you buy bread in bulk or can't finish a loaf, freeze slices or the entire loaf. This prevents it from going stale and allows you to use it when needed.
    - **Store in a cool, dry place**: Keep bread in a breadbox or a paper bag to avoid condensation, which can cause mold growth.
- **Leftovers**:
    - **Cool quickly before refrigerating**: Allow leftovers to cool before placing them in the fridge. This prevents bacterial growth and maintains the food's quality.
    - **Use clear containers**: Using clear storage containers will make it easier to identify what's in your fridge and reduce the likelihood of forgotten leftovers.
- **Dry Goods (grains, pasta, nuts)**:
    - **Use airtight containers**: For grains, pasta, and nuts, airtight containers help prevent moisture and pests from ruining the food. Consider mason jars, vacuum-sealed bags, or sturdy plastic containers.
    - **Store in a cool, dry place**: Pantry items should be kept in a cool, dark area to maintain freshness.

Tip: Always label containers with dates so you can keep track of how long food has been stored and prevent items from being forgotten and going bad.

4.3 Creative Ways to Use Leftovers

Repurposing leftovers is one of the best ways to minimize food waste. Instead of throwing away what's left from dinner, consider how you can transform it into new meals. Here are some creative ways to use up leftover ingredients:

- **Vegetable Scraps**:
  - **Make vegetable broth**: Save the stems, peels, and ends of vegetables like carrots, onions, and celery. Boil them with water and seasonings to create a delicious homemade vegetable broth.
  - **Compost**: If you can't use your veggie scraps, compost them instead of throwing them away to enrich your garden soil.
- **Leftover Cooked Vegetables**:
  - **Add to soups or stews**: Any leftover cooked vegetables can be blended into soups or stews for extra flavor and nutrition.
  - **Vegetable fritters**: Mash up leftover vegetables like mashed potatoes, zucchini, or carrots with some breadcrumbs and eggs to create vegetable fritters.
- **Rice, Pasta, and Grains**:
  - **Stir-fries**: Leftover rice, quinoa, or pasta can be stir-fried with fresh or leftover veggies and protein for a quick meal.
  - **Grain Bowls**: Turn leftover grains into a hearty bowl by adding greens, protein (like beans, chicken, or tofu), and your favorite dressing.
- **Bread**:
  - **Make croutons**: Leftover or stale bread can be cut into cubes, seasoned, and toasted to make croutons for salads or soups.
  - **Bread pudding**: Use stale bread to create a delicious, comforting bread pudding for dessert.
  - **French toast**: Repurpose bread into French toast for a hearty breakfast.
- **Leftover Meat and Protein**:
  - **Wraps or sandwiches**: Leftover meat or roasted vegetables can be placed in a wrap or sandwich with some fresh greens for a quick lunch.
  - **Soup or stew**: Add leftover meats to soups and stews to enhance flavor and nutrition.

Tip: Keep a "leftovers night" once a week where you repurpose any remaining food from the previous days into a new meal. This helps reduce waste and save time cooking.

4.4 Reducing Portion Sizes

Over-portioning is a common cause of food waste. When cooking or serving meals, be mindful of how much food you prepare to avoid throwing away leftovers.

- **Cook smaller portions**: If you're cooking for yourself or your family, prepare only what you need. Start by cooking in smaller quantities and scale up if necessary. It's easier to add more food than to deal with leftovers you can't use.
- **Serve smaller portions**: Try serving smaller portions initially and allow people to go back for more if they're still hungry. This can prevent large amounts of food from being left on plates.
- **Use smaller plates**: Studies have shown that using smaller plates can help reduce portion sizes, leading to less food waste.

4.5 Freezing Leftovers

Freezing leftovers is an excellent way to extend the shelf life of many foods, ensuring that they don't go to waste before they can be eaten.

- **Label frozen foods**: Make sure to label everything you freeze with the date and contents, so you can use them in order of when they were frozen.
- **Freeze in individual portions**: Freeze meals or ingredients in single servings for quick and easy meals later on. This is especially useful for soups, stews, and casseroles.
- **Freeze vegetables**: If you have excess fresh produce that might go bad soon, consider blanching and freezing it for future use. This works especially well for leafy greens, peas, and beans.

4.6 Planning and Shopping Smartly

The best way to reduce food waste starts before you even bring food into your home. Careful planning and smart shopping can help prevent over-purchasing, which is one of the leading causes of waste.

- **Meal Planning**: Plan your meals ahead of time and create a shopping list based on the ingredients you'll need. This way, you can avoid buying food you won't use and reduce the chances of it going bad.
- **Buy in smaller quantities**: For fresh produce, it's better to buy smaller amounts more frequently, rather than purchasing large quantities that might spoil before you can use them.
- **Practice FIFO (First In, First Out)**: When you store food, use the FIFO method—consume older items before newer ones. This helps you

keep track of what needs to be used up first and avoid forgetting food in the back of the fridge or pantry.

Key Takeaways

- **Proper storage**: Store food correctly to extend its freshness and prevent spoilage.
- **Repurpose leftovers**: Use creative recipes to turn leftovers into new meals, reducing food waste.
- **Reduce portion sizes**: Be mindful of how much food you prepare and serve to avoid overproduction and waste.
- **Freeze leftovers**: Freeze extra meals or ingredients to save them for later use.
- **Smart shopping**: Plan meals, buy only what you need, and use FIFO to reduce food waste from the outset.

By following these practices, you can reduce food waste in your home, save money, and minimize your environmental impact. Remember, small changes in how we store, plan, and use food can lead to significant reductions in waste over time.

Chapter 4: Energy Conservation at Home

1. Energy Use and Its Environmental Impact

Energy use is one of the largest contributors to global greenhouse gas emissions and climate change. Every time we turn on a light, use an appliance, or heat our homes, we consume energy—often from non-renewable sources like coal, natural gas, or oil. The environmental impact of this energy use can be profound, as the burning of fossil fuels to generate electricity is a major source of carbon emissions, air pollution, and other harmful effects on the planet.

In this section, we'll dive into how everyday energy consumption contributes to climate change, and why reducing energy use is critical for a sustainable future.

1.1 The Link Between Energy Use and Climate Change

The main source of energy for most homes today comes from the burning of fossil fuels, such as coal, oil, and natural gas. These non-renewable resources release carbon dioxide (CO_2) and other greenhouse gases into the atmosphere, which trap heat and cause the planet's temperature to rise—a phenomenon known as global warming.

- **Carbon Emissions**: When fossil fuels are burned to generate electricity or power vehicles, they release CO_2 into the atmosphere. This excess CO_2 is the main driver of climate change, as it contributes to the greenhouse effect, which raises global temperatures.
- **Air Pollution**: Burning fossil fuels also produces pollutants like sulfur dioxide (SO_2) and nitrogen oxides (NOx), which contribute to smog and acid rain, further damaging the environment and public health.
- **Energy Inefficiency**: Many homes and buildings are not energy-efficient, meaning they waste large amounts of energy. This can increase the demand for electricity from fossil fuel sources, exacerbating the environmental impact.

Takeaway: Every time we use energy derived from fossil fuels, we're contributing to climate change and other environmental problems. Reducing our energy consumption and shifting to cleaner energy sources is crucial to mitigate these impacts.

1.2 The Environmental Impact of Household Energy Use

While much of the energy we use comes from power plants, our everyday household activities play a significant role in our overall energy consumption.

- **Heating and Cooling**: According to the U.S. Department of Energy, nearly half of the energy used in a typical home goes toward heating and cooling. Traditional heating methods, like oil or gas furnaces, as well as air conditioning units that run on electricity, often use fossil fuels and contribute to high energy consumption.
- **Lighting**: Lighting is another major source of household energy use. Incandescent bulbs use significantly more energy than modern LED or CFL bulbs, which offer the same level of brightness but with far less energy consumption.
- **Appliances and Electronics**: Devices like refrigerators, washing machines, and electronics (TVs, computers, etc.) use substantial amounts of energy. Many of these appliances are left on or plugged in when not in use, wasting energy in the form of "phantom" power draw.
- **Water Heating**: Water heaters, especially those that use gas or electricity, consume large amounts of energy. The demand for hot water throughout the day—whether for showers, washing dishes, or laundry—can lead to high energy bills and increased environmental impact.

Takeaway: Common household activities such as heating, lighting, and using appliances are major sources of energy consumption and contribute significantly to carbon emissions and environmental harm. Making your home more energy-efficient can greatly reduce this impact.

1.3 The Need for Energy Conservation

Energy conservation is a key strategy for reducing our environmental footprint and combating climate change. By using less energy in our homes, we can lower greenhouse gas emissions, reduce reliance on fossil fuels, and even save money on energy bills.

- **Decreasing Carbon Footprint**: Reducing energy use helps decrease the amount of CO_2 released into the atmosphere, directly addressing the issue of climate change. By using renewable energy sources, such as solar or wind, or cutting down on our energy consumption, we can reduce our carbon footprint.
- **Conserving Resources**: Fossil fuels, which are finite resources, will eventually run out if we continue using them at current rates. By

reducing energy use and shifting to renewable sources, we can help preserve these resources for future generations.
- **Reducing Pollution**: Decreasing energy demand reduces the need for power plants to burn fossil fuels, thus cutting down on air pollution and its harmful effects on human health and ecosystems.
- **Saving Money**: Many energy-saving measures, such as upgrading to energy-efficient appliances, using programmable thermostats, or installing better insulation, not only reduce environmental harm but also save money in the long run.

Takeaway: Energy conservation is essential for mitigating climate change, preserving natural resources, reducing pollution, and saving money. It's a win-win for both the planet and your wallet.

1.4 The Benefits of Energy Conservation for You and the Planet

Energy conservation offers a host of benefits, both for the environment and for your personal well-being. By making simple changes in how we use energy, we can have a direct impact on the planet's future.

- **Improved Health**: Reducing air pollution from fossil fuel energy sources can lead to improved air quality and better health outcomes. Fossil fuel emissions are linked to respiratory illnesses, heart disease, and other health issues, especially in urban areas.
- **Energy Independence**: By conserving energy and investing in renewable energy sources, you can reduce your reliance on external energy providers, contributing to greater energy independence and stability.
- **Sustainable Future**: Every small change in energy use adds up to a more sustainable future. By conserving energy today, we are helping create a world where future generations can enjoy clean air, clean water, and a stable climate.

Takeaway: Energy conservation not only helps combat climate change but also improves air quality, boosts energy independence, and creates a sustainable future for all.

1.5 How Can We Reduce Our Energy Use at Home?

Now that we understand the environmental impact of energy use, it's time to explore practical ways to reduce our consumption. By taking simple steps at home, we can make a huge difference in lowering our carbon footprints and conserving resources.

- **Switch to Energy-Efficient Appliances**: One of the easiest ways to reduce household energy use is by upgrading to energy-efficient appliances. Look for products with the **Energy Star** label, which signifies that the appliance meets energy efficiency standards set by the government.
- **Seal Leaks and Insulate**: Ensuring that your home is well-insulated and free of drafts can significantly reduce the energy needed for heating and cooling. Weatherproofing windows and doors can save up to 25% of heating and cooling energy.
- **Use Smart Thermostats**: Installing a smart thermostat allows you to set heating and cooling schedules, ensuring that energy is not wasted when you're not at home or when you're asleep.
- **Switch to LED Lighting**: Replace incandescent bulbs with LED or CFL bulbs. These use up to 75% less energy and last much longer, reducing both your energy bill and waste.
- **Unplug Devices**: Many appliances and electronics consume energy even when turned off, a phenomenon known as "phantom load." Unplugging devices when not in use or using a power strip to easily turn off multiple electronics at once can reduce energy waste.
- **Adopt Renewable Energy**: If possible, consider installing solar panels or subscribing to green energy programs offered by your utility company. Solar energy can provide your home with a renewable and sustainable power source.

Takeaway: Simple upgrades and lifestyle changes can dramatically reduce your home's energy consumption, save money, and decrease your environmental impact.

Key Takeaways

- **Energy use is a major contributor** to climate change due to the burning of fossil fuels, which releases greenhouse gases and pollutants.
- **Reducing energy consumption** through energy-efficient appliances, proper insulation, and smarter heating/cooling can lower your carbon footprint and save money.
- **Shifting to renewable energy** sources like solar or wind power further reduces dependence on fossil fuels and helps protect the planet for future generations.
- **Every small change counts**, from switching to LED bulbs to unplugging devices—these actions reduce energy waste and contribute to a more sustainable lifestyle.

By understanding how our energy use contributes to climate change and taking steps to conserve energy, we can all play a role in building a cleaner, greener future.

2. Simple Tips to Save Electricity

Saving electricity doesn't always require expensive upgrades or major changes to your lifestyle. In fact, there are many small, easy-to-implement actions that can significantly reduce your energy consumption without requiring much effort. In this section, we'll explore simple tips and habits you can adopt to save electricity and lower your energy bills, all while contributing to a more sustainable future.

2.1 Turn Off Lights When Not in Use

One of the simplest and most effective ways to reduce electricity consumption is by turning off lights when you leave a room. It may seem obvious, but many people leave lights on unnecessarily when they're not needed.

- **Tip**: Make it a habit to turn off lights as you leave a room, and consider installing timers or motion sensors in areas like hallways and bathrooms to automate this process.
- **Upgrade**: Switch to energy-efficient LED bulbs that use less power and last longer than traditional incandescent bulbs.

Takeaway: Turning off lights when not in use is an easy way to save electricity and extend the lifespan of your light bulbs.

2. Use Natural Light Whenever Possible

Maximizing the use of natural light can drastically reduce your need for artificial lighting during the day. By adjusting your daily activities around natural daylight, you can save electricity and create a brighter, more pleasant home environment.

- **Tip**: Open curtains or blinds during the day to let sunlight in, especially in the morning and afternoon. This reduces the need for artificial lighting.
- **Design Tip**: When decorating or designing a room, use light-colored walls and reflective surfaces to enhance natural light and distribute it more evenly throughout the room.

Takeaway: Relying on natural light reduces your electricity use, improves your well-being, and creates a warm, inviting atmosphere.

2. Unplug Devices When Not in Use

Many electronics and appliances continue to draw electricity even when they are turned off, a phenomenon known as "phantom load." Devices like TVs, computers, chargers, and kitchen appliances are often left plugged in, consuming energy unnecessarily.

- **Tip**: Unplug devices and chargers when they are not in use. If unplugging each device is inconvenient, use a power strip to easily switch off multiple items at once.
- **Power Strips**: Consider using smart power strips that can detect when a device is turned off and automatically cut power to prevent energy waste.

Takeaway: Unplugging unused electronics or using power strips is an easy way to prevent wasted energy and save electricity.

2. Adjust Your Thermostat

Heating and cooling typically account for the largest portion of a household's energy bill. By making small adjustments to your thermostat settings, you can significantly reduce your energy consumption without sacrificing comfort.

- **Tip**: In the winter, set your thermostat to 68°F (20°C) when you're at home and awake, and lower it while you're asleep or away. In the summer, set it to 78°F (26°C) while at home and higher when you're not.
- **Upgrade**: Consider installing a programmable or smart thermostat that can automatically adjust the temperature based on your schedule, so you're not heating or cooling an empty house.

Takeaway: Adjusting your thermostat and investing in a programmable model can reduce energy consumption while maintaining a comfortable living space.

2. Use Energy-Efficient Appliances

Upgrading to energy-efficient appliances is an investment that can pay off in the long run. While the initial cost may be higher, energy-efficient models consume less power, leading to significant savings on your electricity bill over time.

- **Tip**: When purchasing new appliances (such as refrigerators, washing machines, or dishwashers), look for the **Energy Star** label, which indicates the product meets energy efficiency standards.
- **Quick Fix**: If you can't afford new appliances, consider replacing old, inefficient light bulbs with LEDs or replacing aging appliances with newer, more efficient models.

Takeaway: Replacing older appliances with energy-efficient models is a smart, long-term investment that saves both electricity and money.

2. Use the Right Size Appliance for the Job

Using the right size appliance for your needs can help save electricity by avoiding unnecessary energy consumption.

- **Tip**: Use a smaller appliance like a toaster oven or microwave instead of a large oven for small meals. Similarly, choose the appropriate-sized stovetop burner for the pot or pan you're using to avoid wasting heat and energy.
- **Laundry Tip**: Only run your washing machine or dryer with full loads. If you have smaller loads, consider air-drying clothes when possible.

Takeaway: Choosing the right-sized appliance for your tasks prevents excess energy use and helps you avoid wasting electricity.

2. Maintain Your Appliances Regularly

Regular maintenance of your appliances can ensure they are running efficiently and consuming the least amount of electricity possible. Many appliances, such as refrigerators and air conditioners, need occasional cleaning and servicing to operate optimally.

- **Tip**: Clean filters in your air conditioners, furnace, and refrigerators regularly to improve their efficiency. Vacuuming refrigerator coils can reduce energy usage by helping the unit cool more efficiently.
- **Laundry Tip**: Clean your dryer's lint filter after every load to help it run more efficiently and reduce drying time.

Takeaway: Keeping your appliances clean and well-maintained ensures they operate at peak efficiency and use less energy.

2. Opt for Cold Water in Laundry

Washing clothes with hot water consumes a lot of energy, especially when done frequently. By switching to cold water, you can significantly reduce the energy used for laundry without sacrificing cleanliness.

- **Tip**: Use cold water for washing clothes, and adjust your detergent to be effective in lower temperatures.
- **Wash Full Loads**: Only run full loads to maximize efficiency and reduce the number of washes needed.

Takeaway: Washing clothes in cold water reduces energy consumption and is just as effective as hot water for most loads.

2. Seal Gaps and Insulate Your Home

Drafts and poor insulation can lead to your heating and cooling systems working overtime, which consumes more electricity. Sealing gaps and upgrading insulation can keep your home more comfortable while reducing energy costs.

- **Tip**: Check for gaps around windows and doors, and seal them with weatherstripping or caulk to prevent air leaks.
- **Upgrade**: Insulate your home, especially in areas like attics and basements, to keep your home cooler in the summer and warmer in the winter.

Takeaway: Sealing gaps and improving insulation makes your home more energy-efficient, reducing the need for heating and cooling.

2. Use Smart Power Devices

Smart devices can help you save electricity by providing better control over how and when energy is used in your home.

- **Tip**: Invest in **smart plugs, smart thermostats**, and **smart lights** that allow you to control when devices are on or off. These can be scheduled to turn off automatically or controlled via smartphone apps, saving you energy without needing to think about it.
- **Energy Monitor**: Use an energy monitor to track your household's electricity use, identify high-energy appliances, and take steps to reduce unnecessary consumption.

Takeaway: Smart devices can automate and optimize energy use, helping you save electricity without extra effort.

Key Takeaways

- **Turn off lights** when not in use and use natural light to reduce reliance on artificial lighting.
- **Unplug devices** and use power strips to prevent phantom load.
- **Adjust your thermostat** to save energy on heating and cooling, or invest in a programmable thermostat.
- **Choose energy-efficient appliances** and use appliances wisely to minimize energy consumption.
- **Maintain appliances** regularly to ensure they operate efficiently.
- **Opt for cold water** when washing clothes and only run full loads to save energy.
- **Seal gaps** and improve insulation to keep your home more energy-efficient.
- **Use smart devices** to automate and control energy usage in your home.

By incorporating these simple tips into your daily routine, you can significantly reduce your electricity consumption, lower your utility bills, and make a positive impact on the environment. Every small change adds up to a more sustainable lifestyle.

3. Efficient Heating and Cooling: How to Maintain Comfort While Conserving Energy

Heating and cooling are typically the largest sources of energy consumption in the home, accounting for about 50% of energy use in an average household. However, there are numerous ways to maintain comfort while conserving energy. By adopting energy-efficient strategies and making small adjustments to your heating and cooling habits, you can create a more comfortable living environment without wasting energy or increasing your utility bills.

3.1 Understanding the Basics of Heating and Cooling

Before diving into specific tips, it's important to understand how heating and cooling systems work, as well as how they impact your home's energy efficiency.

- **Heating**: Most homes rely on central heating systems that use energy to warm the air or water and circulate it through the home. Common types include gas, electric, and oil furnaces, as well as heat pumps and boilers.

- **Cooling**: Air conditioners, either central or window-mounted, cool the air by removing heat and humidity. They use significant amounts of electricity, especially during peak summer months.
- **Energy Consumption**: Both heating and cooling systems are typically energy-intensive. The higher the temperature differential between inside and outside, the harder the system has to work, leading to increased energy consumption.

3.2 Set Your Thermostat to Optimal Temperatures

One of the easiest ways to save energy on heating and cooling is by setting your thermostat to energy-efficient temperatures.

- **Winter Settings**: In the winter, set your thermostat to **68°F (20°C)** while you are awake and at home. Lower it by a few degrees while you sleep or when you're away—around **60–65°F (15–18°C)** can be sufficient for nighttime or when you're not at home.
- **Summer Settings**: In the summer, keep your thermostat around **78°F (25°C)** when you're at home and active. When you're away or asleep, you can raise it to **85°F (29°C)**, or use fans to keep things comfortable.
- **Programmable Thermostat**: Invest in a **programmable thermostat** to automate these temperature adjustments. Set it to automatically lower heating or cooling when you're not at home, and raise or lower temperatures when you're preparing to wake up or arrive home.

Takeaway: Setting your thermostat to optimal temperatures during different times of day can save significant amounts of energy and money.

3.3 Use Ceiling Fans to Supplement Heating and Cooling

Ceiling fans are an efficient way to help maintain comfort without cranking up the air conditioning or heating. By using fans strategically, you can distribute air more evenly, reducing your reliance on your HVAC system.

- **Winter**: In colder months, set the fan to rotate clockwise on a low speed. This will push warm air that rises back down into the living space, helping to keep the room warmer without raising the thermostat setting.
- **Summer**: During the warmer months, set the fan to rotate counterclockwise to create a wind-chill effect that helps cool the air around you. This can make you feel cooler without lowering the temperature on your thermostat.

- **Tip**: If you don't have ceiling fans, consider using floor fans or box fans to circulate air, especially in bedrooms, to help with cooling.

Takeaway: Ceiling fans (or other types of fans) can enhance comfort by circulating air, reducing the need to use air conditioning or heating excessively.

3.4 Insulate and Seal Your Home

Poor insulation and air leaks can cause significant heat loss in the winter and cool air loss in the summer. By addressing these issues, you can reduce the workload on your heating and cooling systems and keep your home more comfortable year-round.

- **Seal Gaps**: Check for drafts around windows, doors, and other openings. Use weatherstripping or caulk to seal cracks and gaps, which can prevent conditioned air from escaping.
- **Insulation**: Insulate your home, especially the attic and walls, to reduce heat transfer. Proper insulation helps maintain a more stable indoor temperature, making your heating and cooling systems more efficient.
- **Window Insulation**: If you live in an area with extreme temperatures, consider installing **thermal curtains** or **blinds** that can help keep warm or cool air inside. Alternatively, **double-glazed windows** are highly effective at insulating your home.

Takeaway: Sealing air leaks and adding insulation is one of the most cost-effective ways to reduce energy loss and make heating and cooling more efficient.

3.5 Upgrade to Energy-Efficient Heating and Cooling Systems

If your heating or cooling system is outdated, it may be consuming more energy than necessary. Upgrading to a more energy-efficient system can significantly reduce your energy consumption, lower your bills, and reduce your environmental impact.

- **Energy-Efficient Heating**: Consider upgrading to a **high-efficiency furnace** or **heat pump**. Modern heat pumps are particularly energy-efficient because they move heat instead of generating it, which requires less energy than traditional methods.
- **Energy-Efficient Cooling**: If your air conditioner is old, consider replacing it with an **Energy Star-certified** model that uses less electricity. Some modern air conditioners also come with smart

features, such as temperature sensors or automatic humidity control, which make them more efficient.
- **Heat Pumps**: A heat pump can be an excellent solution for both heating and cooling, offering the convenience of one system year-round. They are much more energy-efficient than separate heating and cooling systems.

Takeaway: Upgrading to an energy-efficient heating or cooling system will pay off in the long run by significantly reducing your energy bills.

3.6 Use Zoning and Smart Control Systems

Zoning and smart controls allow you to heat or cool specific areas of your home instead of treating the entire house as one unit. This is especially useful in larger homes, where you may not need to heat or cool every room.

- **Zoning Systems**: If you have a central heating or cooling system, consider installing a **zoning system** that lets you control the temperature in different parts of your home. This allows you to avoid wasting energy on rooms that are not in use.
- **Smart Thermostats**: A **smart thermostat** can learn your habits and adjust the temperature based on when you're home, when you sleep, and when you leave. It can also be controlled remotely via smartphone, so you can adjust the temperature before you get home.

Takeaway: Zoning and smart control systems help you optimize heating and cooling by focusing energy use where and when it's needed, reducing waste and energy consumption.

3.7 Perform Regular Maintenance on Your HVAC System

Routine maintenance is essential to keep your heating and cooling systems running efficiently. Over time, dirt and wear-and-tear can reduce system performance, leading to higher energy consumption.

- **Change Filters Regularly**: Replace or clean air filters every 1-3 months. Clogged filters can reduce airflow, making your heating and cooling system work harder and use more energy.
- **Schedule Annual Inspections**: Have a professional HVAC technician perform an annual inspection of your system. Regular tune-ups can extend the lifespan of your unit and ensure it's operating at peak efficiency.

- **Clean Coils**: If you have a central air conditioner, cleaning the evaporator and condenser coils can improve cooling efficiency and reduce energy consumption.

Takeaway: Regular HVAC maintenance ensures your system runs efficiently, reducing energy waste and prolonging its lifespan.

Key Takeaways

- **Set your thermostat** to energy-efficient temperatures (68°F/20°C in winter, 78°F/26°C in summer).
- **Use ceiling fans** to help circulate air, reducing reliance on heating and cooling systems.
- **Seal air leaks** and ensure your home is well-insulated to prevent heat or cool air from escaping.
- **Upgrade to energy-efficient systems** like high-efficiency furnaces, heat pumps, or Energy Star-rated air conditioners.
- **Use zoning and smart controls** to only heat or cool the areas of your home that are in use.
- **Perform regular maintenance** on your HVAC system to ensure it's operating efficiently and effectively.

By implementing these efficient heating and cooling strategies, you can maintain comfort in your home while significantly reducing energy consumption. These changes not only save you money but also help lower your carbon footprint, contributing to a more sustainable lifestyle.

4. Renewable Energy Options: Exploring Solar, Wind, and Other Renewable Energy Sources

As concerns about climate change and energy sustainability continue to rise, the shift towards renewable energy has become more urgent. Unlike fossil fuels, renewable energy sources are abundant, sustainable, and produce little to no greenhouse gas emissions. In this chapter, we'll explore several renewable energy options, including solar, wind, and other alternative energy sources, to help you make informed decisions about integrating renewable energy into your home and lifestyle.

4.1 Understanding Renewable Energy

Renewable energy refers to energy derived from natural resources that are replenished on a human timescale, such as sunlight, wind, rain, and geothermal heat. These energy sources are considered environmentally friendly because they produce minimal pollution and have a much smaller environmental impact compared to traditional energy sources like coal, oil, and natural gas.

- **Solar Energy**: Captures sunlight and converts it into electricity using photovoltaic panels or into heat using solar thermal systems.
- **Wind Energy**: Converts the kinetic energy of wind into electricity through wind turbines.
- **Other Renewable Energy Sources**: Includes geothermal energy (heat from the Earth), hydropower (energy from flowing water), and biomass (organic materials used as fuel).

4.2 Solar Energy: Harnessing the Power of the Sun

Solar energy is one of the most popular and accessible renewable energy options for homeowners. It uses photovoltaic (PV) panels to convert sunlight into electricity, which can be used to power your home or even be stored for later use. Solar power is clean, abundant, and, once installed, free.

- **Solar Panels**: Photovoltaic (PV) panels are made up of semiconductor materials (typically silicon) that generate electricity when exposed to sunlight. They can be installed on rooftops, or in larger fields, to generate electricity.
    - **Tip**: If you live in an area with abundant sunlight, solar panels are a great investment that can reduce your electricity bills and provide clean energy.
- **Cost of Solar Panels**: The upfront cost of installing solar panels can be high, but federal and state incentives, tax credits, and rebates can significantly reduce the cost. Additionally, many solar providers offer financing options to make the installation more affordable.
- **Solar Battery Storage**: Solar batteries store excess energy produced during sunny days so that it can be used when the sun is not shining. This makes solar energy a more reliable option for off-grid living or for reducing reliance on the grid.
- **Benefits of Solar Energy**:
    - Lowers electricity bills by reducing the need to purchase power from the grid.
    - Reduces your carbon footprint by using clean, renewable energy.
    - Increases property value due to the long-term energy savings.
    - Provides energy independence, especially with battery storage.

Takeaway: Solar power is a clean, renewable energy source that can significantly reduce your energy bills and environmental impact. With the right incentives, it is a cost-effective solution for many homeowners.

4.3 Wind Energy: Harnessing the Power of Wind

Wind energy has long been used as a renewable resource for generating electricity. Wind turbines capture the energy from wind and convert it into electricity. While wind energy is most commonly associated with large-scale wind farms, it is also an option for homeowners who have the space and appropriate conditions.

- **Wind Turbines**: Small residential wind turbines can be installed on your property if you live in an area with consistent and strong winds. These turbines typically generate between 400 watts to 10 kW of power depending on their size.
- **Considerations for Wind Power**:
    - **Wind Speed**: Wind turbines require a certain average wind speed to be effective (usually around 10–12 mph). Areas with consistent wind, such as coastal regions or open plains, are ideal locations for wind power.
    - **Space Requirements**: Wind turbines need ample space to function efficiently. A minimum of one acre of land is recommended to ensure there is enough room for the turbine to operate without interference.
- **Benefits of Wind Energy**:
    - Produces clean, renewable energy with no emissions.
    - Can reduce your electricity bills and provide energy independence.
    - Especially beneficial in areas with consistent winds, such as coastal or open-field locations.

Takeaway: Wind energy can be an effective option if you have the right location, space, and wind conditions. It's a great way to generate renewable energy for your home.

4.4 Geothermal Energy: Using Heat from the Earth

Geothermal energy uses the natural heat from the Earth's core to provide heating and cooling for homes and businesses. It is an increasingly popular renewable energy option, particularly in areas with high geothermal activity, such as the Pacific Northwest and parts of the Midwest.

- **Geothermal Heat Pumps**: These systems tap into the stable temperatures of the Earth's surface to heat and cool your home. In winter, the system draws heat from the Earth, while in summer, it pulls cool air from below the ground.
    - **Installation**: Installing a geothermal heat pump requires drilling into the ground to access the Earth's heat. This can be expensive initially but offers long-term savings through lower energy costs and low maintenance needs.
- **Benefits of Geothermal Energy**:
    - Provides consistent, renewable heating and cooling.
    - Highly energy-efficient with low operating costs once installed.
    - Low environmental impact, as it doesn't rely on burning fossil fuels.

Takeaway: Geothermal energy is a reliable and sustainable option for heating and cooling homes, although it requires a significant upfront investment.

4.5 Hydropower: Harnessing Energy from Water

Hydropower (or hydroelectric energy) is one of the oldest and most well-established forms of renewable energy. It uses the movement of water—such as rivers, waterfalls, or tidal flows—to generate electricity.

- **Small-Scale Hydropower**: Homeowners with access to running water, such as those living near a stream or river, may be able to install a small-scale hydroelectric system. These systems typically consist of a turbine or water wheel that turns as water flows past it, generating electricity.
- **Considerations for Hydropower**:
    - Access to a water source is required.
    - The initial installation cost can be high, especially for systems that require infrastructure such as dams or pipelines.
- **Benefits of Hydropower**:
    - Reliable and consistent energy source, especially in areas with running water.
    - Can generate large amounts of power depending on the size and flow of the water source.

Takeaway: Hydropower can be an excellent renewable energy option for those with access to running water, but it may not be feasible for every homeowner due to geographic and financial considerations.

4.6 Biomass Energy: Turning Organic Waste into Power

Biomass energy involves burning organic materials, such as wood, agricultural waste, or even algae, to generate heat or electricity. This renewable energy source can also be used to produce biofuels, which can power vehicles or machinery.

- **Biomass for Heating**: Some homeowners use wood stoves, pellet stoves, or biomass boilers to heat their homes by burning wood, pellets, or agricultural waste. Biomass heating systems are often more sustainable than traditional fossil fuel-based heating systems.
- **Biomass for Electricity**: Biomass can also be used to generate electricity through combustion or fermentation, where organic material is burned to create steam that drives turbines.
- **Benefits of Biomass**:
    - Reduces reliance on fossil fuels.
    - Can help reduce waste by using agricultural and organic byproducts.
    - Can provide local, sustainable heating solutions.

Takeaway: Biomass energy is a versatile renewable energy option that can be used for both heating and electricity generation, particularly in rural areas where organic materials are abundant.

4.7 The Future of Renewable Energy: Innovation and Integration

As technology advances, the cost of renewable energy continues to fall, making it more accessible to homeowners. The future of renewable energy holds exciting potential with innovations like energy storage, smart grids, and hybrid systems that combine different renewable energy sources for maximum efficiency.

- **Energy Storage**: Improvements in battery technology will allow homes to store renewable energy for later use, making solar and wind energy more reliable.
- **Smart Grids**: Smart grids are electrical networks that can detect and respond to changes in energy demand, allowing homes to seamlessly integrate renewable energy with the power grid.

Takeaway: Renewable energy is rapidly evolving, and the integration of new technologies will make it easier and more affordable for homeowners to adopt these solutions.

Key Takeaways

- **Solar Energy**: Solar panels are a cost-effective and accessible way to generate renewable energy, especially in sunny areas.
- **Wind Energy**: Wind turbines are ideal for locations with strong, consistent winds, but require space and investment.
- **Geothermal Energy**: Geothermal systems provide reliable, renewable heating and cooling but require a significant upfront investment.
- **Hydropower**: Small-scale hydropower systems can be a great option for those with access to a water source, providing consistent energy.
- **Biomass Energy**: Biomass can be used for heating and power generation, utilizing organic waste products.
- **The Future**: Technological advancements in energy storage and smart grids will make renewable energy more accessible and efficient in the future.

By exploring and adopting renewable energy options, you can significantly reduce your carbon footprint, lower your energy bills, and contribute to a more sustainable future. Renewable energy is the key to reducing our dependence on fossil fuels and combating climate change, and there has never been a better time to get started.

Chapter 5: Water Conservation Made Easy

1. The Importance of Water Conservation

Water is one of the most precious resources on Earth, essential for life, agriculture, industry, and daily human activities. However, despite its importance, clean, fresh water is not as abundant as we may think. Water scarcity is a growing global issue, affecting millions of people and ecosystems worldwide. By understanding the significance of water conservation, we can all play a role in preserving this vital resource for future generations.

1.1 What is Water Scarcity?

Water scarcity occurs when the demand for fresh water exceeds the available supply in a region, or when water is too polluted to be usable. According to the United Nations, nearly **two-thirds of the world's population** experiences water stress at least one month per year. In some areas, this shortage is exacerbated by factors such as climate change, population growth, and pollution.

There are two main types of water scarcity:

- **Physical Scarcity**: Occurs when a region does not have enough natural freshwater to meet its needs. This can be due to geographical, environmental, or seasonal factors.
- **Economic Scarcity**: Happens when the water supply is available but inaccessible due to poor infrastructure, financial constraints, or political issues.

Both types of scarcity lead to serious consequences, including the depletion of groundwater supplies, reduced agricultural productivity, and limited access to safe drinking water. This impacts not only human health but also the environment, as ecosystems rely on a steady and clean supply of water to function properly.

1.2 The Global Implications of Water Scarcity

Water scarcity is not just an issue in developing countries—it affects both urban and rural areas in developed nations too. Globally, **approximately 2 billion people** live in countries experiencing high water stress, and about **4 billion people** face severe water shortages during at least one month of the year.

The consequences of water scarcity extend beyond human use. Agriculture, which accounts for about **70%** of global freshwater usage, suffers when there isn't enough water to irrigate crops, leading to food shortages. Reduced freshwater supplies also impact sanitation, hygiene, and the ability to produce clean energy, resulting in a vicious cycle of economic and social instability.

Local Impacts:

- **Rural communities** in water-scarce regions may walk long distances to access clean water, which can take hours each day, limiting time for work, school, and economic activities.
- **Ecosystem degradation** occurs when rivers, lakes, and wetlands dry up, leading to the loss of biodiversity and the collapse of vital ecosystems.
- **Health risks** rise as untreated water becomes more accessible, leading to increased waterborne diseases.

1.3 Why Conserving Water is Crucial

Conserving water is essential for ensuring that we can meet the needs of future generations. By reducing our water usage, we can help alleviate the pressure on water supplies, preserve ecosystems, and contribute to the global effort to address climate change.

Environmental Benefits:

- **Preserves Ecosystems**: When we conserve water, we help maintain healthy wetlands, rivers, and lakes, which are critical for biodiversity.
- **Reduces Energy Consumption**: Water treatment and distribution require energy, and the more water we use, the more energy is needed to process and deliver it.
- **Helps Combat Climate Change**: Efficient water use can help mitigate the impacts of climate change by reducing the energy required for water extraction, treatment, and transport.

Personal and Societal Benefits:

- **Lower Bills**: Water conservation often results in lower water bills, as we consume less water both directly (through showers, washing machines) and indirectly (in food production and energy use).
- **Better Resource Management**: Conserving water helps ensure that local water sources are replenished and available for future use, improving the resilience of communities and industries.
- **Healthier Communities**: With less strain on freshwater supplies, there is a greater opportunity to maintain clean and safe water for drinking, sanitation, and hygiene.

1.4 How Water Conservation Impacts You Personally

Water conservation may seem like an issue that only affects regions facing severe water shortages, but it has a direct impact on your life, no matter where you live. Every action you take to reduce your water consumption, whether it's using less water for daily activities or making sustainable choices in your home and garden, helps to ease the global water crisis and benefits both the environment and your local community.

By conserving water, you are:

- **Reducing your environmental footprint**: Every drop saved contributes to reducing the pressure on natural water resources and ecosystems.
- **Saving money**: By using water more efficiently, you can lower your utility bills.
- **Teaching future generations**: Leading by example helps to instill the importance of water conservation in your children and peers, encouraging sustainable behavior that can have a long-lasting impact.

1.5 The Connection Between Water Conservation and Sustainable Living

Water conservation is a critical component of a sustainable lifestyle. Just as we reduce waste, minimize our energy consumption, and make mindful food choices, it's essential to adopt water-saving habits that align with our overall eco-friendly goals. Small changes—such as fixing leaky faucets, choosing water-efficient appliances, or collecting rainwater for outdoor use—can make a big difference when practiced collectively.

Conserving water also means:

- **Supporting eco-friendly technologies**: Installing water-efficient fixtures like low-flow faucets, showerheads, and dual-flush toilets, or investing in a water-efficient irrigation system for your garden.
- **Engaging in conscious consumption**: Reducing water-intensive products like bottled water and opting for products made with less water (such as organic and locally sourced food).
- **Making better choices at home and in the workplace**: Encouraging workplaces to install water-saving technologies, promote water-efficient practices, and reduce water wastage in daily operations.

By integrating water conservation practices into your life, you are taking a stand for the health of the planet and contributing to the global movement towards sustainable living.

Key Takeaways

- **Water scarcity** is a global challenge that affects billions of people and ecosystems around the world, with serious environmental and economic consequences.
- **Conserving water** is essential to protect our future water supply, preserve ecosystems, and combat climate change.
- **Small actions** in your daily routine—like fixing leaks, reducing shower time, or using water-efficient appliances—can have a significant impact on water conservation.
- **Water conservation** is a key element of sustainable living, and everyone can contribute by making mindful choices and adopting habits that reduce water consumption.

In the following sections, we'll explore practical and easy-to-implement strategies for reducing water usage in your home and daily life. These changes may seem small, but collectively, they can lead to substantial savings and environmental benefits, helping to create a more sustainable future for all.

2. How to Reduce Water Use at Home

Reducing water consumption at home is one of the most effective ways to contribute to water conservation. By adopting simple, practical habits in key areas of your household, you can significantly decrease your water usage without sacrificing comfort or convenience. This chapter provides actionable tips for saving water in three main areas: the kitchen, bathroom, and garden. Let's dive into these practical steps.

2.1 Saving Water in the Kitchen

The kitchen is one of the highest water-using areas in your home, especially when it comes to cooking, cleaning, and dishwashing. By making small adjustments, you can easily cut down your water consumption while still maintaining a functional and efficient kitchen.

2.1.1 Fix Leaks and Drips

Even a small leak in the kitchen can waste a significant amount of water over time. Leaky faucets, pipes, and dishwashers can contribute to gallons of water being lost every day.

- **Action Step**: Regularly check faucets, pipes, and dishwashers for leaks. Repairing a leaky faucet can save you up to **3,000 gallons** of water per year.

2.1.2 Use a Dishwasher Efficiently

Dishwashers use a considerable amount of water, but modern models are designed to be much more water-efficient than washing dishes by hand.

- **Action Step**: Run your dishwasher only when it's full to maximize water efficiency.
- **Action Step**: Choose a dishwasher with the **Energy Star** label, which indicates that it uses less water and energy.

2.1.3 Practice Smart Cooking Techniques

The way we cook can also impact how much water we use.

- **Action Step**: Use a **pressure cooker** or a **slow cooker** to reduce cooking times and water usage.
- **Action Step**: Use a **pot lid** to speed up cooking and reduce the amount of water required.
- **Action Step**: Wash fruits and vegetables in a bowl of water rather than under running water. This prevents unnecessary water waste and can be reused for watering plants later.

2.1.4 Efficiently Wash Dishes

If you prefer to hand-wash dishes, there are a few techniques to reduce water use:

- **Action Step**: Fill one basin with soapy water for washing and another with clean water for rinsing, rather than letting the water run continuously.
- **Action Step**: If washing dishes in the sink, turn off the tap while scrubbing.

2.2 Saving Water in the Bathroom

The bathroom is another area where a significant amount of water is used, especially during showers, baths, and toilet flushing. By being mindful of how we use water in the bathroom, we can make a big impact on overall water consumption.

2.2.1 Install Water-Efficient Fixtures

Modern fixtures can drastically reduce the amount of water used in the bathroom, making a big difference in daily water consumption.

- **Action Step**: Install **low-flow faucets** and showerheads. These fixtures use up to **50% less water** without compromising water pressure.
- **Action Step**: Choose a **dual-flush toilet**, which allows you to select the appropriate flush volume, saving water with each use.

2.2.2 Take Shorter Showers

Showers are among the largest water users in the household. By shortening the length of your showers, you can reduce water consumption dramatically.

- **Action Step**: Aim for **4-5 minute showers** instead of longer ones. This can save you more than **1,000 gallons** of water per year.
- **Action Step**: Install a **shower timer** to track how long you've been in the shower and help curb unnecessary water waste.

2.2.3 Turn Off the Water While Brushing Teeth or Shaving

Leaving the tap running while brushing your teeth or shaving wastes a surprising amount of water.

- **Action Step**: Turn off the water while brushing your teeth or shaving. This small change can save **several gallons** of water each day.

2.2.4 Fix Toilet Leaks

Toilets are notorious for wasting water if they aren't functioning properly.

- **Action Step**: Check your toilet for leaks by adding a few drops of food coloring to the tank and waiting 15 minutes. If the color appears in the bowl, there is a leak.
- **Action Step**: Replace old toilet flappers to stop leaks and prevent water wastage.

2.3 Saving Water in the Garden

Watering the garden is another area where a lot of water is used, particularly in warmer climates. By adopting water-efficient practices in your yard, you can maintain a beautiful garden while saving a significant amount of water.

2.3.1 Water in the Early Morning or Late Evening

Watering your garden at the right time of day ensures that the water is absorbed efficiently and doesn't evaporate quickly due to heat.

- **Action Step**: Water your plants in the early morning or late evening to reduce water loss due to evaporation.
- **Action Step**: Avoid watering during the hottest part of the day, typically between 10 a.m. and 4 p.m., when evaporation rates are highest.

2.3.2 Use a Water-Efficient Irrigation System

Automatic sprinklers can be inefficient if not used properly, often watering areas that don't need it or wasting water due to over-spraying.

- **Action Step**: Install **drip irrigation systems** or **soaker hoses** that deliver water directly to the roots of plants, reducing waste.
- **Action Step**: Use a **rain barrel** to collect rainwater for garden irrigation. This reduces the need for tap water.

2.3.3 Choose Native or Drought-Resistant Plants

The plants you choose for your garden can have a huge impact on how much water you need to use.

- **Action Step**: Opt for **native plants** or **drought-resistant plants**, which are better adapted to local climate conditions and require less water to thrive.
- **Action Step**: Group plants together based on their water needs to ensure efficient watering.

2.3.4 Mulch Your Garden

Mulching helps retain moisture in the soil, reducing the frequency of watering and protecting plant roots from heat.

- **Action Step**: Apply **organic mulch** around your plants. It helps prevent soil from drying out quickly and reduces water evaporation.

2.4 Additional Tips for Water Conservation at Home

Beyond the kitchen, bathroom, and garden, there are other small steps you can take to conserve water throughout your home:

- **Action Step**: Install a **water meter** to track your consumption and identify areas where you can save.
- **Action Step**: Wash full loads of laundry instead of partial loads to maximize water efficiency in the washing machine.
- **Action Step**: Consider using a **water-saving washing machine** that uses less water per load.

Key Takeaways

- **Small actions** can lead to significant water savings across your household. Fixing leaks, installing water-efficient fixtures, and being mindful of your water use can make a big difference.
- **Water-saving appliances** like low-flow faucets, dual-flush toilets, and water-efficient dishwashers can drastically reduce water consumption with minimal effort.
- **In the garden**, water-efficient irrigation systems, proper timing, and choosing the right plants can all help you save water without sacrificing your garden's beauty.
- Every drop counts—by making small, sustainable changes at home, you can contribute to the global effort of conserving water for future generations.

In the next chapters, we'll continue exploring how small eco-friendly actions in other areas of your life can have a positive impact on the planet.

3. Collecting and Reusing Water

One of the most effective ways to reduce your water consumption at home is by collecting and reusing water. By capturing rainwater and repurposing wastewater, you can lower your reliance on municipal water systems and save money on your water bill. This chapter will explore techniques for collecting rainwater, reusing greywater, and other innovative ways to conserve water through reuse.

3.1 Rainwater Harvesting

Rainwater harvesting involves collecting rainwater from roofs or other surfaces and storing it for later use. This practice not only reduces the amount of freshwater you use but also helps to manage stormwater runoff, which can pollute rivers and oceans.

3.1.1 How Rainwater Harvesting Works

The basic principle of rainwater harvesting is simple: when it rains, water falls on your roof, is channeled through gutters and downspouts, and is then collected in a storage tank or barrel. This collected rainwater can be used for various non-potable purposes such as watering plants, cleaning, or flushing toilets.

3.1.2 Setting Up a Rainwater Harvesting System

To start harvesting rainwater, you'll need a few essential components:

1. **Gutters and Downspouts**: Ensure that your gutters are clean and free of debris, allowing water to flow smoothly into your collection system.
2. **First-Flush Diverter**: A first-flush diverter is a device that ensures the first few liters of rainwater, which may contain dust, leaves, and other contaminants, are diverted away from your storage system. This helps ensure the collected rainwater is cleaner.
3. **Storage Tank or Barrel**: The most common storage systems are rain barrels or large tanks. These are typically connected to the downspout with a hose or pipe to collect the water.
4. **Filtration System**: A basic filtration system, such as a mesh screen or sand filter, helps to remove larger debris and keep the water clean.
5. **Pump and Hoses**: To make use of the collected rainwater, you may need a pump and hoses to transfer the water to where it is needed (e.g., garden, outdoor cleaning, etc.).

3.1.3 Benefits of Rainwater Harvesting

- **Reduces Water Bills**: By using harvested rainwater for non-potable uses, you can significantly reduce your reliance on municipal water and lower your water bill.
- **Conserves Freshwater**: Rainwater is a renewable resource, and using it conserves valuable freshwater for drinking, bathing, and cooking.
- **Reduces Stormwater Runoff**: Collecting rainwater reduces the amount of water that runs off your property, which can cause erosion, flooding, and pollution.
- **Improves Plant Growth**: Rainwater is often preferred by plants over tap water because it's free from chemicals like chlorine and fluoride, promoting healthier plant growth.

3.1.4 How Much Water Can You Harvest?

The amount of rainwater you can collect depends on several factors, including the size of your roof, the amount of rainfall in your area, and the size of your storage system.

- **Formula for Estimating Harvested Water**:
  You can estimate the potential amount of rainwater you can collect by using this formula:

  $$\text{Water Collected} = \text{Roof Area (sq ft)} \times \text{Rainfall (inches)} \times 0.623$$

  The result gives you the amount of rainwater in gallons that can be collected for each inch of rainfall.

3.2 Greywater Recycling

Greywater is wastewater from sinks, showers, bathtubs, washing machines, and dishwashers. It does not include water from toilets (which is called blackwater) and is typically safer to reuse for purposes such as irrigation and toilet flushing.

3.2.1 How Greywater Recycling Works

The process of greywater recycling involves diverting wastewater from your household drains, filtering it, and then using it for non-potable purposes. A simple greywater system includes:

1. **Collection**: Water from sinks, showers, and washing machines is collected through plumbing pipes that are rerouted to a storage tank or filtration system.
2. **Filtration**: Basic filtration can remove larger particles, hair, and soap residue. Some systems use natural filters like sand, gravel, or charcoal to improve water quality.
3. **Storage**: Greywater is stored in a safe, clean container until it is ready to be used. It's important to use the water within a few days to avoid bacterial growth.
4. **Distribution**: Greywater is typically used for irrigation or flushing toilets. Some systems include a pump to distribute the water, while others may use gravity.

3.2.2 Benefits of Greywater Recycling

- **Reduces Water Use**: Reusing greywater for non-potable purposes like irrigation can reduce household water consumption by up to **40%**.
- **Decreases Water Bills**: Reusing water for activities like garden watering or toilet flushing can significantly lower your water bill.
- **Minimizes Pollution**: Greywater systems reduce the strain on sewage systems, which helps reduce pollution and conserve fresh water resources.
- **Sustainable Gardening**: Greywater, especially from washing machines (without harsh detergents), can be used to water drought-resistant plants or gardens, enriching the soil with nutrients.

3.2.3 How to Recycle Greywater Safely

While greywater is relatively safe, it's important to ensure that it's properly filtered and used in a way that prevents contamination or health risks:

- **Avoid Harsh Chemicals**: Use eco-friendly cleaning products to ensure that the greywater remains safe for use in your garden. Avoid detergents with phosphates, chlorine, or bleach.
- **Use Greywater Only for Irrigation**: Avoid using greywater on edible plants unless you are sure the system is properly filtered and safe for food crops.
- **Limit Storage Time**: Greywater should not be stored for extended periods, as it can develop harmful bacteria and odors. Use it within **24-48 hours** to maintain freshness.

3.3 Other Water-Saving Techniques

In addition to rainwater harvesting and greywater recycling, there are other ways to reduce water use and waste at home.

3.3.1 Water-Efficient Appliances

Investing in water-efficient appliances can make a significant difference in your home's overall water usage.

- **Action Step**: Choose a washing machine or dishwasher with the **Energy Star** label, which indicates they use less water and energy.
- **Action Step**: Install **low-flow showerheads** and **faucet aerators** to reduce water flow without sacrificing pressure.

3.3.2 Water Conservation in the Garden

As mentioned in previous chapters, watering your garden can be a large source of water waste. Some additional techniques for saving water include:

- **Action Step**: Use a **rainwater irrigation system** to directly water your garden, reducing the need for tap water.
- **Action Step**: **Mulch your plants** to help retain moisture in the soil and reduce the need for frequent watering.

3.3.3 Greywater for Landscaping

For homes with larger gardens, greywater can be used in landscaping projects like **irrigating shrubs**, lawns, and trees. Just be mindful of the types of cleaning agents you use to avoid harming plants.

Key Takeaways

- **Rainwater harvesting** and **greywater recycling** are two of the most effective methods for reducing household water use and conserving valuable resources.

- With a simple **rainwater collection system**, you can capture rainwater for irrigation, cleaning, or toilet flushing.
- **Greywater systems** allow you to reuse wastewater for non-potable purposes, further reducing reliance on fresh water.
- Using water-efficient appliances, practicing mindful water use, and adopting sustainable gardening practices can all contribute to water conservation in your home.

By incorporating these water-saving techniques into your daily routine, you can significantly reduce your water consumption and make a meaningful contribution to water conservation efforts at both the local and global level. In the next chapter, we will continue exploring how to integrate sustainable practices into every aspect of your life.

Chapter 6: Eco-Friendly Transportation

1. Transportation and Carbon Footprint

Transportation is one of the largest contributors to global greenhouse gas emissions. The way we travel—whether by car, bus, airplane, or train—has a significant impact on the environment, contributing to air pollution, climate change, and resource depletion. In this chapter, we will explore how our travel habits affect the planet, the role of transportation in our overall carbon footprint, and strategies for making our transportation choices more eco-friendly.

1.1 Understanding the Carbon Footprint of Transportation

A carbon footprint is the total amount of greenhouse gases (GHGs) emitted into the atmosphere as a result of human activities, typically measured in tons of CO2-equivalent (CO2e) emissions. The transportation sector is responsible for about **14%** of global greenhouse gas emissions, with road vehicles contributing to the largest share.

1.1.1 The Impact of Cars

Personal cars are one of the most significant contributors to the carbon footprint, especially those running on gasoline or diesel. The emissions come from two main sources:

1. **Fuel Combustion**: Cars that run on fossil fuels release carbon dioxide (CO_2), the primary greenhouse gas that contributes to global warming, as well as other pollutants such as nitrogen oxides (NOx) and particulate matter (PM).
2. **Manufacturing and Maintenance**: The production of cars, including the extraction of raw materials, manufacturing processes, and vehicle maintenance (such as tire replacement and oil changes), all contribute additional environmental impacts.

A typical gasoline-powered car emits approximately **4.6 metric tons** of CO2 per year, depending on driving habits and fuel efficiency.

1.1.2 Air Travel and Its Environmental Impact

Airplanes are known for their high carbon emissions due to the amount of fuel required to operate them. The aviation industry accounts for about **2-3%** of global CO2 emissions, with long-haul flights having a disproportionate impact. The environmental cost of air travel is significant, as airplanes burn large amounts of fuel per passenger, particularly during takeoff and landing.

A **round-trip flight from New York to London** can emit about **1.6 tons of CO2** per passenger. This is roughly equivalent to the annual emissions of driving a car for a year in some regions.

1.1.3 Public Transportation: A Greener Alternative

Public transportation, including buses, trains, and subways, is generally much more eco-friendly than driving a personal car or flying. These modes of transport are typically more efficient and have a lower carbon footprint per passenger.

For example:

- **A bus** can carry dozens of passengers, making it far more energy-efficient than each person driving their own car.
- **Trains**, especially electric ones, are among the least carbon-intensive options, especially in regions that use renewable energy sources for electricity generation.

In cities with well-developed public transportation networks, using these services rather than personal vehicles can significantly reduce your carbon footprint.

1.2 Strategies for Reducing the Environmental Impact of Transportation

While we may not be able to completely eliminate our need to travel, there are many ways to reduce the environmental impact of our transportation choices. In this section, we'll explore practical steps to make transportation more sustainable.

1.2.1 Drive Less: The Power of Walking and Cycling

One of the simplest ways to reduce your carbon footprint is to avoid driving whenever possible. Opting for walking or cycling instead of driving for short trips not only cuts emissions but also benefits your health.

- **Action Step**: For trips under **2 miles**, choose to walk or bike instead of driving. This is often quicker and more eco-friendly, especially in urban areas.
- **Action Step**: If you live in a walkable neighborhood or have easy access to cycling paths, consider incorporating more active transportation into your daily routine.

1.2.2 Carpool and Ride-Sharing

Carpooling with others or using ride-sharing services is an effective way to reduce the number of vehicles on the road, thereby lowering overall emissions.

- **Action Step**: If you commute to work or school, explore carpooling with others who live nearby or have similar schedules. This can halve the carbon footprint of each individual's commute.
- **Action Step**: Use ride-sharing apps like **Uber** or **Lyft** in combination with public transportation, or coordinate carpools with friends, family, or colleagues.

1.2.3 Choose an Eco-Friendly Vehicle

If driving is unavoidable, opting for a more fuel-efficient or electric vehicle can drastically reduce your carbon footprint.

- **Action Step**: Consider purchasing a **hybrid** or **electric vehicle (EV)**. Electric vehicles produce zero tailpipe emissions, and hybrid vehicles combine gasoline and electric power to achieve much higher fuel efficiency.
- **Action Step**: Look for **fuel-efficient cars** that get better mileage. Choosing a vehicle with good fuel economy reduces the amount of fuel you consume and the emissions you produce.

1.2.4 Use Public Transportation

Where possible, use **public transportation**—such as buses, trains, subways, or trams—especially in cities with extensive networks.

- **Action Step**: Opt for public transportation instead of driving whenever possible, particularly for commuting, longer trips, or traveling to crowded city centers.
- **Action Step**: In areas where public transportation is less accessible, explore alternatives like **shared bicycle services**, **car-sharing** options, or **electric scooters** for shorter distances.

1.2.5 Fly Less, Travel Smarter

If you travel frequently for business or leisure, reducing the number of flights you take can make a huge difference in your overall carbon footprint.

- **Action Step**: Whenever possible, **take the train** instead of flying for domestic or short-haul trips. High-speed trains, particularly in Europe and Asia, are an excellent low-carbon alternative to flying.

- **Action Step**: **Combine trips** or opt for **virtual meetings** if your travel is for business. This eliminates the need for flying and reduces both time and environmental impact.
- **Action Step**: When flying is necessary, **choose direct flights** over connecting ones. Takeoff and landing are the most fuel-intensive parts of a flight, so fewer takeoffs result in less overall emissions.

1.3 The Benefits of Eco-Friendly Transportation

Switching to more sustainable transportation options offers several benefits beyond just reducing your carbon footprint. Here are some of the advantages:

- **Cleaner Air**: Fewer cars on the road means less air pollution, leading to improved air quality and better public health.
- **Reduced Traffic Congestion**: Carpooling, public transportation, and cycling help reduce traffic congestion, leading to faster, more efficient travel for everyone.
- **Lower Costs**: While public transportation and cycling can be cost-effective alternatives, driving a fuel-efficient car or electric vehicle also saves money in the long run due to lower fuel costs and fewer maintenance requirements.
- **Improved Health**: Walking and cycling not only reduce emissions but also improve physical health by increasing daily activity levels.
- **Less Noise Pollution**: Fewer cars mean a quieter environment, contributing to a more peaceful urban and suburban landscape.

Key Takeaways

- **Transportation is a major contributor** to global carbon emissions, with cars, air travel, and other modes of transport impacting the environment significantly.
- **Making eco-friendly transportation choices**—such as walking, cycling, carpooling, using public transit, and driving fuel-efficient or electric vehicles—can reduce your carbon footprint.
- **Flying less** and choosing alternatives like trains or virtual meetings can help reduce the environmental impact of air travel.
- By adopting more sustainable transportation habits, you not only help the environment but also enjoy health benefits, cost savings, and a better quality of life.

In the next chapters, we will explore additional sustainable living practices, focusing on waste management, eco-friendly products, and green building practices.

2. Choosing Greener Ways to Travel

As we become more aware of the environmental impact of our travel choices, adopting greener transportation options has never been more important. The good news is that there are many alternatives to driving alone or flying that can significantly reduce our carbon footprint while still allowing us to move efficiently from one place to another. This chapter will explore four eco-friendly ways to travel: **walking, biking, carpooling,** and **public transportation**.

2.1 Walking: The Ultimate Green Transportation

Walking is the most sustainable mode of transportation available. It's zero-emissions, free, and offers numerous health benefits. Whether it's walking to the corner store or for a leisurely stroll, this simple activity can be incorporated into daily life in many ways.

2.1.1 Benefits of Walking

- **Zero Emissions**: Walking produces no carbon emissions, making it the greenest transportation option. It's completely energy-efficient, requiring nothing more than human effort.
- **Health Benefits**: Walking promotes physical health by improving cardiovascular fitness, lowering stress levels, and boosting overall well-being.
- **Cost-Free**: There are no costs associated with walking, other than comfortable footwear, making it an incredibly affordable option.
- **Reduces Traffic**: By walking short distances instead of driving, you help reduce congestion, making travel more efficient for everyone.

2.1.2 How to Walk More

- **Action Step**: Start by **walking for short trips**. For errands that are less than 1-2 miles, walking is often faster and more convenient than driving.
- **Action Step**: **Plan your day around walking**. If possible, schedule your activities so that you can walk between destinations rather than using a car.
- **Action Step**: Invest in a **comfortable pair of shoes** that you can wear while walking long distances.
- **Action Step**: Walk with a purpose. Instead of driving to the gym, consider walking to exercise, or use walking as a form of stress relief during your lunch break.

2.2 Biking: A Fast, Green, and Fun Alternative

Biking is another highly eco-friendly mode of transportation. Like walking, it is human-powered and produces no emissions. It's faster than walking and is ideal for medium-distance travel. Biking also offers health benefits and is often quicker than driving in urban areas with heavy traffic.

2.2.1 Benefits of Biking

- **Zero Emissions**: Bicycles are entirely powered by you, meaning they produce no pollution.
- **Physical Exercise**: Cycling strengthens the heart, muscles, and bones while improving flexibility and endurance.
- **Cost-Effective**: Biking has a low cost of entry—after an initial investment in a bike, maintenance is minimal. Additionally, you avoid the costs associated with fuel and parking.
- **Reduced Traffic and Parking Hassles**: Biking allows you to avoid traffic jams and expensive parking spaces in busy urban areas.
- **Accessibility**: Cycling infrastructure, such as bike lanes and rental programs, is growing in many cities, making it easier than ever to bike.

2.2.2 How to Bike More

- **Action Step: Incorporate biking into your daily commute**. If your workplace or school is within biking distance, consider switching from driving or public transport to cycling.
- **Action Step**: Take advantage of **bike-sharing programs** in cities that offer them. You don't need to own a bike to use it—simply rent one when needed.
- **Action Step**: Invest in a **quality bike lock** and a safety helmet for protection.
- **Action Step**: Explore **bike-friendly routes**. Many cities have mapped out cycling lanes or trails for safer and more enjoyable biking experiences.

2.3 Carpooling: Sharing the Ride

Carpooling is an effective way to reduce the number of vehicles on the road, cutting down on both carbon emissions and traffic congestion. By sharing rides with others, you lower your individual environmental impact and reduce costs associated with fuel and parking.

2.3.1 Benefits of Carpooling

- **Reduced Carbon Emissions**: Carpooling reduces the number of cars on the road, decreasing traffic and the amount of pollution generated per trip.
- **Cost Savings**: By sharing the cost of fuel and parking, carpooling is a great way to save money.
- **Decreased Traffic Congestion**: Fewer cars mean less traffic, making the commute faster and less stressful for everyone.
- **Increased Social Interaction**: Carpooling can provide an opportunity to network or catch up with friends and colleagues.

2.3.2 How to Carpool More

- **Action Step**: Organize **carpooling with colleagues** or friends who have similar work hours and live nearby. Apps like **Waze Carpool** and **Splt** can help match you with carpool partners.
- **Action Step**: If carpooling with others isn't an option, look for **ride-sharing services** like **Uber Pool** or **Lyft Line**. These services allow multiple passengers to share the same ride, reducing emissions and costs.
- **Action Step**: Take turns driving. This allows each person to share the responsibility of driving while cutting down on emissions.

2.4 Public Transportation: Efficient and Eco-Friendly

Public transportation includes buses, subways, trams, and trains, all of which are more energy-efficient than driving alone. They carry many passengers at once, which means they produce fewer emissions per person compared to single-occupancy vehicles.

2.4.1 Benefits of Public Transportation

- **Lower Emissions**: Public transportation significantly reduces the number of vehicles on the road, resulting in less pollution and fewer greenhouse gas emissions.
- **Cost-Effective**: Public transportation is often much cheaper than owning and maintaining a car, particularly when factoring in fuel, insurance, and parking fees.
- **Convenient and Efficient**: In urban areas, public transportation is often faster than driving, especially during peak traffic times.
- **Supports Sustainable Development**: Public transit reduces urban sprawl, preserves open spaces, and fosters more sustainable city designs.

2.4.2 How to Use Public Transportation More

- **Action Step**: Familiarize yourself with your local **public transit system**, including routes, schedules, and pricing. Many cities also offer apps to make trip planning easier.
- **Action Step**: Invest in a **monthly or annual pass**, which can save you money in the long term and encourage regular use.
- **Action Step**: If possible, combine **public transport with walking or cycling** for the last leg of your journey. This reduces your overall carbon footprint and keeps you active.
- **Action Step**: **Support local transit initiatives** that work to improve the accessibility, efficiency, and sustainability of public transportation options.

Key Takeaways

- **Walking, biking, carpooling, and public transportation** are all sustainable alternatives to driving and flying, offering significant environmental, financial, and health benefits.
- By choosing **greener ways to travel**, we can significantly reduce our carbon footprint, ease traffic congestion, and promote healthier communities.
- These alternatives are often more affordable, more convenient, and better for the environment than relying on private cars.
- Whether you're making small changes in your daily commute or opting for greener travel on longer trips, each decision you make can help create a more sustainable future.

3. Switching to Electric or Hybrid Vehicles

As we look for ways to reduce our carbon footprint, transitioning to electric (EV) or hybrid vehicles is one of the most effective choices. These vehicles offer significant environmental benefits by producing fewer emissions than traditional gasoline or diesel cars. In this chapter, we will explore the benefits and considerations of switching to eco-friendly cars, including electric and hybrid options, and how they can contribute to a more sustainable lifestyle.

3.1 Understanding Electric and Hybrid Vehicles

Before diving into the benefits and considerations, it's important to understand the difference between electric and hybrid vehicles:

- **Electric Vehicles (EVs)**: Fully electric cars are powered entirely by electricity stored in batteries. They run on electric motors, producing zero emissions from the tailpipe. EVs need to be recharged through home charging stations or public charging points.
- **Hybrid Vehicles**: Hybrids combine a gasoline or diesel engine with an electric motor. They can operate on electric power alone for short distances or switch to the gasoline engine when needed. Hybrids are more fuel-efficient than traditional cars and emit fewer emissions.

Both EVs and hybrids are a step forward in reducing reliance on fossil fuels and decreasing air pollution compared to traditional gasoline-powered vehicles.

3.2 Benefits of Electric and Hybrid Vehicles

Switching to electric or hybrid cars offers several key advantages, from environmental to financial benefits.

3.2.1 Environmental Benefits

- **Zero or Low Emissions**: EVs produce no tailpipe emissions, which means they help improve air quality and reduce greenhouse gases (GHGs) that contribute to climate change. Even hybrid vehicles, which still use gasoline, have a significantly lower carbon footprint than traditional cars due to their ability to operate on electric power at times.
- **Reduced Dependence on Fossil Fuels**: Electric vehicles run on electricity, which can be generated from renewable sources like solar or wind. This reduces the demand for oil and other fossil fuels.
- **Cleaner Cities**: By reducing tailpipe emissions, EVs and hybrids help reduce urban air pollution, improving the health of city residents and enhancing quality of life.

3.2.2 Financial Benefits

- **Lower Fuel Costs**: Charging an EV is generally much cheaper than fueling a gasoline-powered car. Even hybrid vehicles, which still use gasoline, achieve better fuel economy, lowering overall fuel costs.
- **Tax Incentives and Rebates**: Many governments offer **tax credits**, **rebates**, and **incentives** for purchasing electric and hybrid vehicles. These can reduce the initial purchase price and make these vehicles more accessible to consumers.
- **Lower Maintenance Costs**: EVs have fewer moving parts than traditional cars, meaning they generally require less maintenance. There are no oil changes, fewer brake repairs (due to regenerative braking), and fewer mechanical issues overall.
- **Reduced Insurance Costs**: Many insurers offer discounts for electric or hybrid cars due to their lower risk of accidents and their environmentally-friendly image. In some regions, driving an eco-friendly car can also help reduce registration and road tax costs.

3.2.3 Performance and Convenience

- **Smooth and Quiet Ride**: Electric motors are quieter and provide instant torque, resulting in a smooth, quiet, and powerful driving experience. The acceleration in EVs is typically faster and more responsive than in traditional gasoline vehicles.
- **Convenient Charging**: Charging EVs is easier than ever. Many electric car owners charge their vehicles at home using a regular electrical outlet or a dedicated charging station. Public charging infrastructure is also expanding globally, making it easier to charge on the go.
- **Reduced Range Anxiety**: While earlier EV models had limited driving ranges, today's electric vehicles offer ranges comparable to traditional cars, with some models capable of driving over 300 miles on a single charge. The increasing number of charging stations ensures that drivers can find places to charge during long trips.

3.3 Considerations Before Switching to an EV or Hybrid

While the benefits are clear, it's important to weigh a few considerations before making the switch to an electric or hybrid vehicle.

3.3.1 Upfront Cost

- **Higher Initial Price**: EVs and hybrids tend to have a higher upfront cost compared to traditional cars, primarily due to the cost of their batteries. However, this is offset by lower operating costs over the vehicle's lifespan. Government rebates and tax incentives can also help reduce the initial financial burden.
- **Long-Term Savings**: Despite the higher purchase price, many EV owners find that they save money in the long term due to lower fuel and maintenance costs.

3.3.2 Charging Infrastructure

- **Availability of Charging Stations**: While the network of charging stations is growing rapidly, it may still be limited in some areas, especially rural or less-developed regions. Be sure to check the availability of public charging stations near your home, workplace, or frequently visited locations.
- **Home Charging**: If you have a garage or parking space, you can install a Level 2 home charger, which charges the vehicle much faster than a standard outlet. However, if you live in an apartment or lack a private parking space, charging could be more challenging.
- **Charging Time**: Charging an EV can take several hours, depending on the type of charger and the vehicle's battery capacity. However, advancements in fast-charging technology are reducing wait times significantly.

3.3.3 Battery Life and Disposal

- **Battery Longevity**: One of the most important factors to consider when buying an electric vehicle is the lifespan of the battery. EV batteries generally last between **8-15 years**, but their performance may degrade over time. However, many manufacturers offer warranties for their batteries, and the technology is continually improving.
- **Recycling and Disposal**: The disposal of EV and hybrid batteries is an important environmental consideration. Recycling systems for EV batteries are still evolving, but many companies are working on creating sustainable solutions for the recycling and reuse of EV batteries.

3.4 Is an Electric or Hybrid Vehicle Right for You?

Choosing whether to switch to an electric or hybrid vehicle depends on your lifestyle and driving needs. Here are some factors to consider:

- **Driving Distance**: If you often drive long distances, a hybrid may be a better choice since it has a longer range before needing to recharge. However, if your typical daily commute is short, an EV could easily meet your needs.
- **Availability of Charging Stations**: If you live in a place with abundant charging infrastructure, an EV may be very convenient. In areas where charging stations are limited, a hybrid might be more practical.
- **Environmental Priorities**: If reducing your carbon footprint is a top priority, an electric vehicle is the clear winner, as it produces zero tailpipe emissions.
- **Budget**: While EVs may have a higher upfront cost, the long-term savings on fuel and maintenance may make it a financially viable option over time. Additionally, consider government incentives and rebates that can reduce the price.

Key Takeaways

- **Electric and hybrid vehicles** offer significant environmental benefits by producing fewer emissions than traditional gasoline-powered cars.
- **EVs** produce zero emissions and can be powered by renewable energy, while **hybrids** offer reduced fuel consumption and fewer emissions than conventional cars.
- **Financial benefits** of owning an EV or hybrid include lower fuel and maintenance costs, tax incentives, and possible discounts on insurance.
- **Considerations** include the upfront cost, availability of charging infrastructure, and the longevity of EV batteries.
- Switching to an electric or hybrid vehicle is an important step toward reducing your personal carbon footprint and supporting a cleaner, more sustainable future.

4. Reducing Travel Emissions

Traveling, whether for work or leisure, contributes significantly to global carbon emissions. However, by making conscious decisions about how we travel, we can minimize our environmental impact and help reduce the carbon footprint associated with transportation. This section explores practical strategies for reducing travel-related emissions, from choosing greener modes of transportation to offsetting emissions for unavoidable long-distance travel.

4.1 Choose Low-Emission Transportation Options

One of the most effective ways to reduce travel emissions is to prioritize low-emission transportation options whenever possible. Here are a few strategies to consider:

4.1.1 Opt for Public Transportation

Public transportation is far more energy-efficient than private cars because it carries more people per trip, reducing the number of vehicles on the road and lowering overall emissions. Whether it's buses, trains, or subways, these options are usually more environmentally friendly than driving alone.

- **Action Step**: When traveling within a city, use public transport such as buses, trams, or trains instead of taxis or rental cars.
- **Action Step**: In regions with high-quality transit systems, plan your trips to maximize the use of trains, light rail, or long-distance buses.

4.1.2 Fly Less, Fly Smart

Air travel is one of the most significant contributors to travel-related emissions, especially for long-haul flights. Reducing the number of flights you take or flying more efficiently can make a big difference.

- **Action Step**: Whenever possible, **choose alternative forms of transportation** for shorter distances. Trains and buses are often more eco-friendly alternatives to short domestic flights.
- **Action Step**: If flying is necessary, opt for **direct flights** rather than ones with layovers. Direct flights are more fuel-efficient since takeoff and landing are the most energy-intensive phases of flight.
- **Action Step**: Consider **flying with airlines that prioritize sustainability**. Some airlines invest in fuel-efficient planes or biofuels, which can reduce the environmental impact of air travel.

4.1.3 Drive Less and Share the Ride

If public transportation isn't available, reducing the number of car trips you make or opting for shared rides can help lower emissions.

- **Carpool**: Share rides with friends, family, or coworkers whenever possible. Apps like **BlaBlaCar** or **Waze Carpool** make it easy to find people traveling in the same direction.
- **Use Ride-Sharing Services**: Choose ride-sharing services like **Uber Pool** or **Lyft Line**, which allow multiple passengers to share the same trip and reduce the number of vehicles on the road.
- **Action Step**: Consider **renting fuel-efficient vehicles** if you need to drive for longer trips. Hybrid or electric rental cars are becoming more widely available.

4.2 Offset Your Carbon Footprint

When long-distance travel is unavoidable, you can mitigate its environmental impact by purchasing carbon offsets. Carbon offset programs allow travelers to compensate for the emissions generated by their trips by supporting projects that reduce or remove carbon from the atmosphere.

4.2.1 How Carbon Offsetting Works

Carbon offsetting involves calculating the emissions produced by your travel and then purchasing credits to fund projects that reduce emissions elsewhere, such as reforestation, renewable energy, or energy efficiency programs.

- **Action Step**: Use a **carbon footprint calculator** (available online) to estimate the emissions generated by your trip. Many airlines and travel companies offer carbon offset programs during the booking process.
- **Action Step**: Purchase carbon offsets from **reputable providers** that support certified projects, such as **Gold Standard**, **Verified Carbon Standard**, or **Climate Action Reserve**.
- **Action Step**: Support **local carbon offset initiatives** if available, as these often provide social benefits for communities alongside environmental improvements.

4.3 Practice Sustainable Travel Habits

Beyond the mode of transport, how you conduct yourself during travel also matters. By adopting sustainable habits while traveling, you can reduce waste, energy use, and emissions.

4.3.1 Travel Light

The heavier your luggage, the more fuel your vehicle (whether it's a plane, car, or bus) needs to move it. Packing light not only makes traveling easier but also reduces the carbon footprint of your trip.

- **Action Step**: Limit your luggage to the essentials and avoid overpacking. If flying, opt for carry-on luggage whenever possible, which reduces the weight of the aircraft and saves fuel.
- **Action Step**: Choose **multi-functional clothing** and **reusable travel accessories** to minimize the need for disposable items during your trip.

4.3.2 Conserve Energy at Hotels or Rentals

Hotels and accommodations consume significant energy through heating, cooling, lighting, and water use. By making small adjustments to your behavior, you can help reduce the environmental impact of your stay.

- **Action Step**: **Reuse towels and linens** during your hotel stay rather than requesting fresh ones daily. This reduces water and energy use.
- **Action Step**: Turn off lights, air conditioning, or heating when leaving your room, and unplug electronic devices when not in use.
- **Action Step**: Stay at **eco-friendly hotels** or accommodations that have sustainability certifications, such as **LEED**, **Green Key**, or **EarthCheck**. These places typically use energy-efficient appliances, solar power, and water-saving measures.

4.4 Plan Environmentally Friendly Vacations

Sustainable travel isn't just about reducing emissions—it's about choosing experiences that support environmental protection and cultural preservation.

4.4.1 Choose Eco-Tourism Destinations

Eco-tourism focuses on traveling to natural areas that conserve the environment, respect local cultures, and provide economic support to local communities.

- **Action Step**: When planning your vacation, look for **eco-tourism destinations** that promote conservation, sustainable practices, and respect for the environment.
- **Action Step**: Consider booking **eco-friendly tours** that limit their impact on wildlife and fragile ecosystems.

4.4.2 Support Local Economies

Traveling sustainably also means supporting local businesses, which reduces the carbon footprint associated with importing goods and helps foster sustainable economic development.

- **Action Step**: When possible, **support locally-owned restaurants**, shops, and hotels that use sustainable practices. Buying local also reduces the environmental impact of shipping goods from faraway locations.
- **Action Step**: Participate in **community-based tourism**, where locals guide you through their culture and environment, ensuring your visit benefits the local community directly.

4.5 Sustainable Commuting Practices

In addition to reducing emissions while traveling long distances, making small changes to your daily commute can have a significant impact on your overall carbon footprint.

4.5.1 Walking and Biking for Short Distances

As mentioned in the previous chapter, walking or biking for short distances can drastically reduce your emissions. For daily commuting, these options are both environmentally friendly and healthy.

- **Action Step**: Walk or bike for short-distance commutes to work, school, or errands. It's a zero-emissions alternative to driving and can improve your overall well-being.
- **Action Step**: Consider using a **bike-sharing program** in urban areas for quick trips if owning a bike isn't feasible.

Key Takeaways

- **Low-emission transportation options** like public transportation, carpooling, and biking are some of the most effective ways to reduce travel-related emissions.
- **Offsetting carbon emissions** through certified projects helps neutralize the environmental impact of unavoidable air travel or long-distance trips.
- **Sustainable travel habits**—such as packing light, conserving energy in hotels, and supporting eco-tourism—are essential for reducing your environmental footprint while exploring the world.
- By making **conscious travel choices**, you can enjoy new destinations while protecting the environment and supporting local communities.

In the next chapter, we will continue exploring sustainable living by diving into water conservation practices and how to reduce water usage in our homes and daily lives.

Chapter 7: Sustainable Fashion and Consumption

1. The Impact of Fast Fashion

Fast fashion refers to the rapid production of inexpensive clothing to meet ever-changing trends, often at the cost of the environment and labor conditions. In this section, we will explore the environmental and social impact of fast fashion, from resource depletion to pollution and waste, and why shifting toward more sustainable fashion choices is essential.

1.1 The Environmental Cost of Fast Fashion

The fast fashion industry is one of the most environmentally damaging sectors, with a wide-reaching impact on ecosystems and natural resources. Let's break down some of the major environmental issues associated with fast fashion production.

1.1.1 Water Consumption and Pollution

The clothing industry is a major consumer of water, particularly in cotton production and textile dyeing. For example, producing a single cotton T-shirt can require up to **2,700 liters** of water, which is about the amount a person drinks in 2.5 years. This high water demand places significant strain on water-scarce regions where many textiles are produced.

Moreover, textile dyeing is one of the largest contributors to water pollution globally. Factories often discharge untreated wastewater into rivers and oceans, polluting waterways with harmful chemicals, dyes, and heavy metals, which harm aquatic life and contaminate drinking water.

- **Action Step**: Learn about the water footprint of your clothing and support brands that practice sustainable water management and use eco-friendly dyes.

1.1.2 Greenhouse Gas Emissions

The fast fashion industry is a significant source of carbon emissions. Synthetic fabrics like polyester, nylon, and acrylic are derived from fossil fuels, and the manufacturing of these materials releases large amounts of greenhouse gases. Even natural fibers like cotton, though biodegradable, require high amounts of water and pesticides, both of which have indirect environmental costs.

Transporting clothing from factories, often located in developing countries, to consumers worldwide adds to the carbon footprint. Air and sea freight used for transporting fashion products emit substantial amounts of carbon dioxide, further contributing to climate change.

- **Action Step**: Consider buying clothes made from natural, organic, or recycled fibers and reducing the carbon footprint by choosing locally produced garments.

1.1.3 Textile Waste and Landfills

Fast fashion's rapid turnover of trends encourages overconsumption, leading to vast amounts of clothing waste. Millions of tons of textiles end up in landfills each year, where synthetic fabrics take hundreds of years to decompose. As these materials break down, they release **methane**, a potent greenhouse gas, and contribute to soil and water contamination through leaching of chemicals.

Additionally, many garments are made from blended fibers (e.g., polyester-cotton blends), which complicates recycling, further increasing the amount of waste that ends up in landfills.

- **Action Step**: To reduce textile waste, prioritize high-quality, durable clothing that lasts longer and avoid purchasing trendy items that will go out of style quickly.

1.1.4 Microplastic Pollution

Synthetic fabrics, such as polyester, are not only energy-intensive to produce but also contribute to **microplastic pollution**. Every time synthetic clothes are washed, tiny plastic fibers are shed and eventually make their way into rivers, oceans, and the food chain. These microplastics are harmful to marine life and can even end up in the seafood we eat.

- **Action Step**: Reduce your reliance on synthetic fabrics and wash clothes less frequently or use a **microplastic filter** for your washing machine to capture plastic fibers.

1.2 The Social and Ethical Cost of Fast Fashion

Fast fashion is notorious for exploiting cheap labor in developing countries, where workers often face unsafe working conditions, low wages, and long hours. Factories, sometimes referred to as **sweatshops**, operate under poor labor regulations, where workers have little to no protection from labor abuse.

The pressure to produce clothing quickly and cheaply can also lead to tragedies such as the **Rana Plaza collapse** in Bangladesh in 2013, where over a thousand garment workers lost their lives due to unsafe building conditions.

By choosing sustainable fashion, consumers can support ethical production processes that prioritize the well-being of workers and uphold fair trade practices.

- **Action Step**: Look for certifications such as **Fair Trade, Global Organic Textile Standard (GOTS)**, and **B Corporation** when purchasing clothing to ensure ethical labor practices.

1.3 The Role of Consumerism in Fast Fashion

The success of fast fashion is largely driven by consumer demand for cheap, trendy clothing. Fast fashion brands release new collections every few weeks, encouraging consumers to keep buying and discarding clothing. This cycle of overconsumption not only fuels environmental degradation but also places pressure on garment workers to meet high production demands under stressful conditions.

Reducing our consumption and opting for more sustainable choices can disrupt this harmful cycle.

- **Action Step**: Practice **mindful consumption** by questioning whether you really need new clothes or if your current wardrobe can be repurposed. When purchasing, invest in quality items that will last longer.

Key Takeaways

- The fast fashion industry has significant negative environmental impacts, including high water consumption, pollution, greenhouse gas emissions, textile waste, and microplastic pollution.
- Fast fashion's labor practices often exploit workers in developing countries, leading to unsafe working conditions and low wages.
- Consumer demand plays a pivotal role in sustaining the fast fashion industry, but by making mindful choices, we can support a shift toward more sustainable, ethical fashion.

In the next section, we will explore strategies for making more sustainable fashion choices and how to build an eco-friendly wardrobe without compromising style.

2. Slow Fashion: Quality Over Quantity

Slow fashion is a movement that advocates for mindful, intentional consumption and production of clothing. Unlike fast fashion, which prioritizes speed and low cost, slow fashion focuses on quality, sustainability, and ethics. In this section, we'll explore the principles of slow fashion, why it matters, and how choosing durable, ethically made clothes can reduce your environmental footprint while supporting responsible businesses.

2.1 What is Slow Fashion?

Slow fashion is about producing clothes with care, using environmentally friendly materials, and ensuring ethical working conditions for those involved in the supply chain. It's the opposite of fast fashion's model of mass production and constant consumption, promoting instead the idea of buying fewer, higher-quality items that last longer and have a smaller ecological impact.

Here are the core values that define slow fashion:

- **Durability**: Slow fashion emphasizes clothing that is built to last, both in terms of material quality and timeless design.
- **Sustainability**: This involves using eco-friendly materials, reducing waste, and minimizing carbon emissions during production.
- **Ethical Production**: Slow fashion brands prioritize fair wages, safe working conditions, and transparency in their supply chains.
- **Mindful Consumption**: Slow fashion encourages consumers to buy only what they need, invest in versatile pieces, and avoid overconsumption.

2.2 Why Choose Quality Over Quantity?

One of the primary principles of slow fashion is prioritizing **quality** over **quantity**. While fast fashion encourages frequent purchases of inexpensive, low-quality items, slow fashion invites consumers to invest in pieces that are well-made and will last for years, reducing the need to constantly replace worn-out or out-of-trend garments.

Here are some reasons why opting for quality over quantity is a sustainable choice:

2.2.1 Durability Saves Resources

High-quality clothing is often made from more durable fabrics and stitched with better craftsmanship, meaning it can withstand the test of time and frequent use. When you invest in well-made garments, they are less likely to fall apart after a few washes, reducing the overall demand for new clothing.

- **Action Step**: Look for clothing made from **natural, durable materials** such as organic cotton, wool, linen, or Tencel. These materials are more resilient and have less environmental impact than synthetic fabrics.
- **Action Step**: Avoid fast fashion trends and focus on building a **capsule wardrobe**—a collection of versatile, timeless pieces that can be mixed and matched for various occasions.

2.2.2 Reducing Waste

The longer a piece of clothing lasts, the less waste it generates. By choosing long-lasting garments, you help decrease the amount of textile waste that ends up in landfills. A well-maintained, durable piece of clothing can serve you for years, whereas a fast fashion item might only last a season before being discarded.

- **Action Step**: Extend the lifespan of your clothing by following care instructions carefully, repairing items when needed, and storing them properly.
- **Action Step**: Embrace **upcycling** and **thrifting**—repurpose older garments into something new or shop second-hand to find quality pieces that have already stood the test of time.

2.2.3 Timeless Style Over Trends

Slow fashion promotes timeless designs that transcend fleeting trends. Fast fashion encourages consumers to constantly chase the latest styles, resulting in a cycle of continuous purchases. In contrast, slow fashion focuses on **classic, versatile pieces** that never go out of style, allowing you to build a wardrobe that is functional, fashionable, and sustainable.

- **Action Step**: Invest in **neutral colors** and **classic cuts** that you can wear year-round. These pieces are less likely to feel outdated, making it easier to avoid buying trendy but short-lived fashion items.
- **Action Step**: Prioritize **versatility**—buy pieces that can be dressed up or down for different occasions, allowing you to get more use out of fewer items.

2.3 Supporting Ethical Brands

A key aspect of slow fashion is supporting brands that prioritize ethical production practices. This means buying from companies that ensure fair wages and safe working conditions for their employees and that actively seek to reduce their environmental footprint.

2.3.1 Transparency in the Supply Chain

Ethical fashion brands are transparent about where and how their clothes are made. They provide information about the materials used, the factories they work with, and the environmental impact of their products. This level of transparency is crucial in ensuring that your clothing choices align with your values.

- **Action Step**: Before buying from a brand, check whether they publish details about their supply chain, including where their clothes are made and what materials are used. Look for certifications like **Fair Trade**, **GOTS**, or **Cradle to Cradle** that indicate ethical practices.
- **Action Step**: Support **local artisans** or smaller brands that focus on handmade, small-batch production. These companies often have more control over their supply chain and prioritize quality and craftsmanship.

2.3.2 Fair Labor Practices

Many fast fashion brands rely on underpaid labor in developing countries to produce cheap clothing quickly. By contrast, slow fashion companies often commit to paying their workers fair wages, providing safe working environments, and offering benefits that improve the well-being of their employees.

- **Action Step**: Look for brands that prioritize **worker rights** and offer **living wages**. Certifications such as **Ethical Trade Initiative (ETI)** or **WRAP** ensure that factories meet international labor standards.
- **Action Step**: Avoid brands with a history of labor violations, even if their products are cheap and trendy. Support companies that empower their workers through fair trade practices.

2.4 Building a Sustainable Wardrobe

Transitioning to slow fashion means making thoughtful, intentional choices about what we buy and how we wear it. A sustainable wardrobe consists of fewer but higher-quality pieces that are versatile and long-lasting.

2.4.1 Invest in Staple Pieces

Rather than buying into every trend, focus on **staple wardrobe pieces**—the essentials that can be styled in various ways. These are often items like well-made jeans, classic blazers, a little black dress, comfortable shoes, and neutral-colored tops.

- **Action Step**: When shopping, think about how many different outfits you can create from a single item. The more versatile a piece, the more likely it is to stay in your wardrobe for years to come.

2.4.2 Care for Your Clothes

Taking good care of your clothing can significantly extend its lifespan, which in turn reduces the need to buy new items. Simple practices like washing your clothes less frequently, using cold water, and air-drying instead of machine drying can help preserve fabric quality.

- **Action Step**: Follow the **care instructions** on your clothing labels to maintain their quality. Invest in tools like fabric shavers to keep knitwear looking fresh, and repair small damages like loose buttons or torn seams rather than discarding the item.
- **Action Step**: Consider **natural fabric care** products that are free of harsh chemicals, which can be better for your clothes and the environment.

Key Takeaways

- Slow fashion promotes quality over quantity, focusing on durability, ethical production, and mindful consumption.
- Investing in long-lasting, timeless pieces not only saves resources but also reduces waste and supports more sustainable fashion choices.
- Supporting ethical brands that prioritize fair labor practices and transparency in their supply chains can make a significant difference in reducing the environmental and social harm caused by fast fashion.
- Building a sustainable wardrobe is about choosing fewer, high-quality items and taking care of them properly, ensuring they last for years rather than months.

In the next section, we will explore how to identify sustainable fabrics and materials, and why your choice of fabric plays a crucial role in the overall impact of your wardrobe.

3. Secondhand and Upcycling: Sustainable Shopping and Giving New Life to Old Items

One of the most effective ways to reduce the environmental impact of your wardrobe is to embrace **secondhand shopping** and **upcycling**. By purchasing pre-loved items or transforming old clothes into something new, you can extend the life cycle of garments, reduce waste, and break away from the fast fashion cycle. In this section, we'll discuss how to shop sustainably through secondhand markets and how to get creative with upcycling to give new life to old clothes.

3.1 The Benefits of Secondhand Shopping

Secondhand shopping, whether at thrift stores, vintage shops, or online platforms, offers a sustainable alternative to buying brand-new clothes. Not only does it reduce demand for the production of new garments, but it also helps keep clothing out of landfills.

3.1.1 Reducing Environmental Impact

When you buy secondhand clothing, you're giving a garment a new lease on life, which reduces the need for new resources like water, energy, and raw materials that would otherwise go into producing new clothing. This is particularly important considering the resource-intensive nature of garment production, especially in the fast fashion industry.

For example, a pair of jeans can require up to **10,000 liters of water** to produce. By purchasing secondhand, you bypass that environmental cost entirely.

- **Action Step**: Visit local thrift stores, consignment shops, and online secondhand platforms like **Depop**, **Poshmark**, or **ThredUp**. You'll often find high-quality, lightly used items at a fraction of the cost of new ones.

3.1.2 Reducing Waste

Secondhand shopping also helps reduce textile waste. Each year, millions of tons of clothing are discarded and end up in landfills. Buying secondhand helps divert clothes from this waste stream and encourages a more circular economy, where items are reused rather than disposed of after a short life.

- **Action Step**: Make it a habit to donate or sell your gently used clothes rather than throwing them away. Platforms like **eBay** or local charity shops can give your unwanted items a second life.

3.1.3 Unique and Vintage Finds

Shopping secondhand often gives you access to unique or vintage pieces that you won't find in mainstream stores. Vintage clothing tends to be higher quality, with timeless designs and durable fabrics. Plus, buying secondhand allows you to express your individuality, as many items are one-of-a-kind or no longer in production.

- **Action Step**: Explore vintage markets and online vintage stores to find high-quality, distinctive items. These pieces are often made from durable materials and designed to last, making them both stylish and sustainable.

3.2 Tips for Successful Secondhand Shopping

Shopping secondhand can be a treasure hunt, but with a few strategies, you can make the most of it and find great sustainable additions to your wardrobe.

3.2.1 Know What You Need

It's easy to get carried away when shopping secondhand, but it's still important to be mindful of your purchases. Before you go shopping, make a list of what you need and stick to it. This prevents impulse buying, which can lead to clutter and overconsumption.

- **Action Step**: Assess your current wardrobe and make note of any gaps or staple pieces you're missing. Whether you're looking for a winter coat or a pair of jeans, having a plan will help you avoid unnecessary purchases.

3.2.2 Inspect for Quality

When buying secondhand, always check the quality of the items. Look for signs of wear and tear, such as holes, stains, or loose stitching, but don't be discouraged by minor issues that can be easily fixed. Well-made garments, especially vintage ones, tend to be more durable and can last for years if properly cared for.

- **Action Step**: Check the fabric, seams, and zippers to ensure they're in good condition. If you find a small flaw, consider whether it can be repaired or altered to your liking.

3.2.3 Be Open to Alterations

Sometimes, the perfect secondhand find might not fit exactly as you'd like. But with a bit of tailoring or DIY alterations, you can customize clothes to suit your style and fit. Tailoring secondhand clothes can often be more affordable than buying new, especially for high-quality or designer items.

- **Action Step**: Don't hesitate to buy an item that's a little too big or long—it's easier to take something in than to let it out. Develop basic sewing skills or find a good tailor who can help you adjust pieces to fit perfectly.

3.3 Upcycling: Turning Old Clothes into New Creations

Upcycling is the process of transforming old, worn-out, or unused clothing into new, functional, or fashionable items. It's a creative and sustainable way to refresh your wardrobe without buying new clothes, and it helps reduce textile waste by giving items a second life.

3.3.1 The Environmental Benefits of Upcycling

Upcycling extends the life of garments that might otherwise end up in landfills, reducing waste and the demand for new resources. It also encourages creativity and personal expression, allowing you to customize your wardrobe in ways that reflect your unique style.

Upcycling can be as simple as altering a T-shirt into a tote bag or as complex as combining fabrics from multiple garments to create a completely new piece.

- **Action Step**: Gather old or worn-out clothes from your wardrobe that you no longer wear. Think creatively about how they can be transformed—an old dress can become a top, or a pair of jeans can be turned into shorts or a bag.

3.3.2 DIY Upcycling Ideas

If you're new to upcycling, start with simple projects and build your skills over time. Here are some easy upcycling ideas to get started:

- **T-shirt Tote Bag**: Turn an old T-shirt into a reusable tote bag by cutting off the sleeves and stitching the bottom shut.
- **Denim Shorts**: Cut an old pair of jeans into shorts, and add patches or embroidery for a personalized touch.
- **Patchwork Clothing**: Use fabric scraps from different items to create a unique patchwork jacket, skirt, or bag.
- **Embroidery**: Add custom embroidery to an old jacket or shirt to give it a fresh, updated look.
- **Action Step**: Experiment with different DIY techniques like cutting, sewing, dyeing, or fabric painting to customize your old clothes. Many online tutorials and upcycling communities can provide inspiration and guidance.

3.3.3 Collaborating with Local Artisans

If DIY isn't your style, consider working with local artisans or seamstresses who specialize in upcycling. Many small businesses focus on transforming old garments into beautiful, one-of-a-kind pieces, often incorporating creative techniques that give the item a new personality.

- **Action Step**: Research local upcycling businesses or artisans in your community. By supporting them, you're contributing to a circular economy and encouraging creative, sustainable fashion practices.

3.4 Creating a Circular Wardrobe

One of the goals of embracing secondhand shopping and upcycling is to move toward a **circular wardrobe**—a collection of clothes that are continuously worn, repaired, and passed on, rather than discarded. By extending the life cycle of garments, we can significantly reduce the demand for new clothing and minimize our environmental impact.

- **Action Step**: Start by identifying pieces in your wardrobe that can be upcycled or donated. Incorporate secondhand shopping into your routine, and when buying new, choose items that are high-quality and designed to last.

Key Takeaways

- Secondhand shopping reduces demand for new resources and helps keep clothing out of landfills, making it a sustainable option for building your wardrobe.
- Upcycling old garments gives them a second life, reduces textile waste, and encourages creativity and personal expression.
- By focusing on secondhand and upcycled fashion, you contribute to a more sustainable, circular fashion system, minimizing the environmental impact of your clothing choices.
- Small repairs, alterations, and creative DIY projects can breathe new life into old clothes, helping you build a unique, sustainable wardrobe without the need for constant new purchases.

In the next section, we'll explore sustainable fabrics and materials, and how your choice of fabric can play a key role in reducing the environmental impact of your wardrobe.

4. Minimalism and Mindful Consumption: Reducing Overall Consumption and Buying with Intention

Adopting a minimalist approach and practicing mindful consumption are essential components of living a more sustainable and eco-friendly lifestyle. Minimalism encourages you to focus on what truly adds value to your life while reducing unnecessary consumption. Mindful consumption emphasizes making intentional choices, buying only what you need, and prioritizing quality and sustainability. In this chapter, we explore how embracing minimalism and mindfulness can help you simplify your life, reduce waste, and make more conscious, sustainable decisions.

4.1 What is Minimalism?

Minimalism is a lifestyle that encourages you to intentionally reduce the clutter—both physical and mental—in your life. It's about focusing on the essentials, eliminating the excess, and prioritizing what truly matters. At its core, minimalism advocates for quality over quantity, mindful decision-making, and a more intentional relationship with material possessions.

- **Action Step**: Begin by decluttering your home, particularly your wardrobe. Sort through your belongings and keep only what you truly use and value. Donate, sell, or repurpose items that no longer serve a purpose in your life.
- **Action Step**: Embrace a **"one in, one out"** rule—whenever you bring a new item into your home, let go of an old one to maintain balance and prevent clutter.

4.1.1 The Environmental Benefits of Minimalism

Minimalism and sustainability go hand in hand. By consuming less, you reduce your environmental footprint, as fewer resources are used to produce, transport, and dispose of products. Additionally, a minimalist mindset encourages mindful consumption, helping you make more sustainable choices and avoid overbuying.

- **Action Step**: Instead of succumbing to trends or impulse buying, focus on **thoughtful purchases**. Ask yourself, "Do I need this?" and "Will this item add value to my life?" before making a purchase.
- **Action Step**: Prioritize **quality over quantity**—invest in fewer, well-made items that are durable, timeless, and versatile, rather than constantly buying cheap, disposable products.

4.2 Practicing Mindful Consumption

Mindful consumption is about making deliberate, thoughtful decisions when it comes to buying and using products. It's about understanding the true impact of your purchases—on both the environment and the people involved in producing them. Mindful consumption also means being aware of your own needs and values, and resisting societal pressure to overconsume.

4.2.1 Understanding the Life Cycle of a Product

Mindful consumption involves considering the entire life cycle of a product, from its production to its disposal. This includes asking questions like:

- Where was this item made, and what resources were used in its production?
- How was it transported, and what is its carbon footprint?
- Will it last a long time, or is it designed to be used briefly and then discarded?
- Can it be recycled or repurposed at the end of its life, or will it contribute to landfill waste?
- **Action Step**: Before buying something new, research the brand's sustainability practices and the materials used in their products. Choose brands that prioritize eco-friendly materials, ethical labor practices, and low environmental impact.

4.2.2 Conscious Decision-Making: Needs vs. Wants

Mindful consumption encourages you to distinguish between **needs** and **wants**. Many of us are conditioned by advertising and societal expectations to constantly desire more, even when we don't truly need it. By being more intentional in your purchases, you can cut down on unnecessary consumption and avoid waste.

- **Action Step**: When considering a purchase, take a moment to reflect on whether it's a true necessity or simply a passing desire. Try the **30-day rule**: if you want something, wait 30 days. If you still feel it's essential after that time, then consider purchasing it.
- **Action Step**: Create a **wish list** for non-essential items. By putting them on the list instead of buying them immediately, you give yourself time to reconsider whether you truly want or need them.

4.3 The Benefits of Minimalism and Mindful Consumption

Adopting minimalism and practicing mindful consumption offer several key benefits, both for the environment and for your personal well-being:

4.3.1 Reducing Waste

When you adopt a minimalist mindset, you naturally reduce the amount of waste you generate. By consuming less and focusing on essentials, you avoid overbuying, which means less packaging, fewer discarded items, and less clutter in your home.

- **Action Step**: Focus on **buying less but better**. Whether it's clothing, household goods, or technology, choose items that are durable and designed to last, rather than trendy or cheap items that will need to be replaced quickly.

4.3.2 Saving Resources

The fewer products you buy, the fewer natural resources are used in manufacturing, shipping, and distributing those goods. Minimalism helps conserve these resources, contributing to a lower environmental impact.

- **Action Step**: Reduce your demand for resource-intensive goods like fast fashion, electronics, and single-use items. Opt for secondhand goods, repair rather than replace, and look for products made from sustainable materials.

4.3.3 Financial Savings

Minimalism isn't just good for the planet; it's also beneficial for your wallet. By reducing unnecessary purchases and focusing on what truly matters, you can save money that might otherwise be spent on impulse buys, unnecessary upgrades, or items that don't bring long-term value.

- **Action Step**: Track your spending for a month and identify areas where you can cut back. You may find that small changes, like avoiding fast fashion or skipping impulse buys, can lead to significant savings over time.

4.3.4 Improved Mental Well-Being

Minimalism promotes clarity, peace, and mental well-being by reducing clutter and the constant desire for more. When you free yourself from the pressure to overconsume, you can focus more on what brings you genuine happiness—whether it's spending time with loved ones, enjoying nature, or engaging in meaningful activities.

- **Action Step**: Embrace **gratitude** for the things you already have. Rather than constantly seeking more, focus on appreciating the items, experiences, and people that enrich your life.

4.4 How to Cultivate Minimalism and Mindful Consumption

Transitioning to a minimalist lifestyle and practicing mindful consumption doesn't happen overnight. It's a gradual process that involves shifting your mindset and habits over time. Here are some practical steps to help you get started:

4.4.1 Declutter and Simplify

Begin your minimalist journey by decluttering your living space. Go through your belongings and keep only the items that bring you joy, serve a purpose, or add value to your life. Let go of the rest—whether through donations, selling, or recycling.

- **Action Step**: Tackle one area of your home at a time, such as your closet, kitchen, or garage. Ask yourself if each item is truly necessary or adds value to your life. If not, consider letting it go.
- **Action Step**: Embrace digital minimalism by decluttering your digital life—unsubscribe from email lists, delete unused apps, and streamline your digital devices to reduce distraction and stress.

4.4.2 Focus on Quality, Not Quantity

When making purchases, prioritize quality over quantity. Choose well-made, durable items that will last a long time, rather than cheap, disposable products that will wear out quickly. This not only reduces waste but also helps you build a collection of meaningful possessions that serve you well over the long term.

- **Action Step**: Invest in items that are multi-functional, durable, and made from sustainable materials. Whether it's clothing, furniture, or technology, choosing quality ensures that your possessions last longer and reduce the need for replacements.

4.4.3 Mindful Shopping Habits

Adopt a mindful approach to shopping by setting intentions before you buy. When you feel the urge to shop, pause and ask yourself why—are you fulfilling a genuine need, or are you responding to external influences like advertising or boredom?

- **Action Step**: Before making a purchase, ask yourself a series of **mindful questions**: Do I really need this? Will it last? Where was it made? How does it align with my values? This practice can help you make more informed, conscious decisions.
- **Action Step**: Try **intentional shopping challenges**, such as a **"no-buy month"** or **"project 333"** (a minimalist wardrobe challenge where you wear only 33 items for 3 months), to help reset your shopping habits and cultivate a deeper appreciation for the things you already own.

Key Takeaways

- Minimalism is about reducing excess and focusing on what truly matters, which leads to less consumption, less waste, and a simpler, more intentional life.
- Mindful consumption encourages you to make deliberate, thoughtful purchasing decisions that consider the environmental and ethical impact of your choices.
- Practicing minimalism and mindful consumption benefits both the planet and your personal well-being, reducing waste, conserving resources, and promoting mental clarity.
- Decluttering, prioritizing quality over quantity, and adopting mindful shopping habits can help you transition to a more sustainable, minimalist lifestyle.

In the next chapter, we'll explore how sustainable living goes beyond the individual and can be applied at a community level, promoting collective environmental action.

Chapter 8: Green Cleaning and Home Care

1. Toxic Chemicals in Household Products: The Dangers of Conventional Cleaning Products

Most conventional household cleaning products contain a wide range of chemicals that can pose significant risks to both human health and the environment. These products, which include everything from surface cleaners and detergents to air fresheners and disinfectants, often contain toxins that contribute to air and water pollution, harm wildlife, and can lead to health issues for those who use them.

1.1 Health Risks of Toxic Chemicals

Many conventional cleaning products contain ingredients such as ammonia, chlorine, phthalates, and volatile organic compounds (VOCs), all of which can have harmful effects on human health. Immediate reactions can include headaches, dizziness, skin irritation, and respiratory problems. Prolonged exposure to certain chemicals has been linked to more serious health conditions like asthma, hormonal imbalances, and even cancer.

- **Volatile Organic Compounds (VOCs)**: These are found in many common cleaning products and are known to evaporate into the air, contributing to indoor air pollution. They can cause respiratory problems and exacerbate conditions like asthma and allergies.
- **Phthalates**: Often used in synthetic fragrances, phthalates are endocrine disruptors, meaning they can interfere with the body's hormone systems and potentially cause reproductive and developmental problems.
- **Ammonia**: Found in many glass and surface cleaners, ammonia can cause skin and respiratory irritation, and when mixed with bleach, it creates a toxic gas that can be extremely dangerous.
- **Chlorine**: Used in disinfectants and bleach products, chlorine exposure can lead to respiratory issues, eye irritation, and skin burns. It is also harmful to aquatic life when it washes down the drain.

1.2 Environmental Impact of Chemical Cleaners

The use of conventional cleaning products has far-reaching environmental consequences. When we use these products, their toxic ingredients are often washed down the drain, entering our water systems. Many of these chemicals are not removed by water treatment plants, which means they can end up in rivers, lakes, and oceans, where they can harm aquatic life and disrupt ecosystems.

- **Water Pollution**: Chemicals such as phosphates, found in many detergents, can lead to water pollution, causing algae blooms that deplete oxygen levels and harm marine life.
- **Air Pollution**: Many cleaning products release harmful chemicals into the air, contributing to indoor and outdoor air pollution. VOCs, for example, contribute to the formation of smog, which has a negative impact on both environmental and public health.
- **Plastic Waste**: Many conventional cleaning products come in single-use plastic packaging, contributing to the global plastic waste crisis. These plastic bottles often end up in landfills or oceans, where they take hundreds of years to decompose and pose a threat to wildlife.

1.3 The Hidden Costs of "Clean"

While conventional cleaning products may promise to make your home clean and fresh, the hidden costs—both to your health and the environment—are substantial. Fortunately, there are safer, greener alternatives available that allow you to maintain a clean home without the risks associated with toxic chemicals. In the next section, we will explore how to make the switch to eco-friendly cleaning products and how to create your own effective, non-toxic cleaners using natural ingredients.

2. Making Your Own Eco-Friendly Cleaners: DIY Recipes for Non-Toxic Cleaning Solutions

Switching to eco-friendly cleaning doesn't have to mean buying expensive products. Many effective, non-toxic cleaners can be made at home using simple, natural ingredients you may already have in your pantry. DIY cleaning solutions not only reduce your exposure to harmful chemicals but also help cut down on plastic packaging waste and lower your overall environmental impact.

Here are some easy-to-make, eco-friendly cleaning recipes that use natural ingredients like vinegar, baking soda, essential oils, and castile soap. These natural cleaners are just as effective as their commercial counterparts but without the harmful side effects.

2.1 Essential Ingredients for Green Cleaning

Before diving into the recipes, it's helpful to understand the key ingredients commonly used in DIY eco-friendly cleaners and why they are so effective:

- **White Vinegar**: A powerful natural disinfectant that cuts through grease and kills bacteria. Its acidic nature helps break down mineral deposits, soap scum, and grime.
- **Baking Soda**: A mild abrasive that is great for scrubbing surfaces. It also neutralizes odors and can be used to tackle tough stains.
- **Castile Soap**: A plant-based soap that is biodegradable and gentle on the skin. It's effective for general cleaning and can be used for dishes, floors, and even laundry.
- **Lemon Juice**: Naturally acidic and antibacterial, lemon juice helps to remove stains, cut grease, and leave surfaces shiny.
- **Essential Oils**: Not only do essential oils add a pleasant scent to your cleaners, but many (like tea tree, lavender, eucalyptus, and lemon) have natural antibacterial, antifungal, and antiviral properties.
- **Hydrogen Peroxide**: A natural disinfectant that is safe for the environment. It's great for cleaning surfaces and removing stains.

2.2 DIY Green Cleaning Recipes

Here are some simple, effective DIY cleaning recipes to get you started:

2.2.1 All-Purpose Cleaner

This basic all-purpose cleaner is perfect for wiping down countertops, cleaning appliances, and disinfecting surfaces in the kitchen and bathroom.

Ingredients:

- 1 cup white vinegar
- 1 cup water
- 10-20 drops of essential oil (optional, for scent)

Instructions:

1. Mix the vinegar and water in a spray bottle.
2. Add the essential oil (if desired) for fragrance and additional antibacterial properties (e.g., tea tree or lavender).
3. Shake well before each use. Spray onto surfaces and wipe clean with a cloth.

2.2.2 Glass and Mirror Cleaner

Get streak-free shine on your windows and mirrors with this simple, non-toxic solution.

Ingredients:

- 1 cup water
- 1 cup white vinegar
- 1 tablespoon cornstarch (optional, for extra streak-fighting power)

Instructions:

1. Combine water, vinegar, and cornstarch in a spray bottle.
2. Shake well to ensure the cornstarch is fully dissolved.
3. Spray on windows or mirrors and wipe with a microfiber cloth or newspaper for a streak-free finish.

2.2.3 Scrubbing Paste (for Tough Stains)

For tougher cleaning jobs, like scrubbing the bathtub, removing grime from tile, or cleaning the oven, this simple scrubbing paste does the trick.

Ingredients:

- ½ cup baking soda
- ¼ cup castile soap
- A few drops of water (as needed)

Instructions:

1. Mix baking soda and castile soap in a small bowl to form a paste.
2. If the mixture is too thick, add a few drops of water until it reaches a spreadable consistency.
3. Apply the paste to the surface, scrub with a sponge or brush, and rinse clean.

2.2.4 Toilet Bowl Cleaner

This natural toilet cleaner will leave your toilet fresh and clean without harsh chemicals.

Ingredients:

- ½ cup baking soda
- ¼ cup white vinegar
- 10 drops tea tree essential oil (optional)

Instructions:

1. Sprinkle baking soda into the toilet bowl.
2. Add the vinegar (it will fizz) and let it sit for 5-10 minutes.
3. Scrub the bowl with a toilet brush, then flush.

4. Add a few drops of tea tree oil for extra disinfecting power if desired.

2.2.5 Floor Cleaner

This gentle but effective cleaner works well on wood, tile, or laminate floors.

Ingredients:

- 1 gallon of warm water
- ½ cup white vinegar
- ¼ cup castile soap
- 10 drops of essential oil (optional, for scent)

Instructions:

1. Combine the ingredients in a bucket.
2. Use a mop to clean floors as usual, making sure to wring out excess water to avoid damaging wood floors.
3. Let the floors air-dry or wipe with a clean cloth for a streak-free finish.

2.2.6 Carpet Deodorizer

If your carpets need a refresh, this natural deodorizer will neutralize odors without harsh chemicals.

Ingredients:

- 1 cup baking soda
- 10 drops essential oil (such as lavender or eucalyptus)

Instructions:

1. Mix the baking soda and essential oil in a jar or shaker.
2. Sprinkle the mixture evenly over carpets.
3. Let it sit for at least 15 minutes (longer if possible).
4. Vacuum thoroughly to remove the baking soda and odors.

2.2.7 Wood Polish

For polishing wood furniture, this simple recipe adds shine and protects the surface.

Ingredients:

- ½ cup olive oil
- ¼ cup white vinegar
- 10 drops lemon essential oil (optional, for scent)

Instructions:

1. Combine the olive oil and vinegar in a spray bottle or bowl.
2. Add lemon essential oil for fragrance and additional cleaning power.
3. Apply a small amount to a soft cloth and buff the wood in circular motions.

2.3 Benefits of DIY Eco-Friendly Cleaners

Making your own cleaning products has numerous benefits, both for your health and the environment:

- **Non-Toxic**: These DIY recipes avoid the harmful chemicals found in many commercial cleaners, reducing your exposure to toxins and improving indoor air quality.
- **Environmentally Friendly**: Using natural ingredients means fewer chemicals are washed down the drain, helping to reduce water pollution and environmental damage.
- **Cost-Effective**: Many DIY cleaners use inexpensive household ingredients, saving you money in the long run.
- **Plastic-Free**: By making your own cleaners, you can reuse containers and reduce the amount of single-use plastic bottles you would otherwise purchase.

2.4 Customizing Your DIY Cleaners

One of the great advantages of making your own cleaning solutions is the ability to customize them to suit your preferences. You can experiment with different essential oils to find scents you love or adjust the strength of your cleaners based on your needs.

- **Scent**: Try lavender for a calming effect, eucalyptus for a fresh, invigorating scent, or lemon for a bright, clean fragrance.
- **Strength**: For more stubborn grime, you can increase the concentration of vinegar or baking soda. For lighter cleaning tasks, you can dilute your solutions with more water.

Key Takeaways

- DIY eco-friendly cleaners are effective, affordable, and easy to make, using simple, natural ingredients like vinegar, baking soda, and essential oils.

- By making your own cleaners, you can reduce your exposure to harmful chemicals, protect the environment, and save money.
- These non-toxic solutions are just as powerful as conventional cleaners, making them a smart, sustainable choice for your home.

In the next section, we'll explore how to create a toxin-free laundry routine, including natural detergents and fabric softeners.

3. Sustainable Home Maintenance: How to Keep Your Home in Good Shape While Reducing Your Environmental Impact

Maintaining a home involves more than just cleaning—it requires regular upkeep and repairs to ensure everything functions efficiently. Sustainable home maintenance focuses on minimizing your environmental impact while keeping your home in good shape. By using eco-friendly materials, energy-efficient tools, and sustainable practices, you can reduce waste, conserve resources, and even save money on utility bills.

This section will guide you through practical steps and tips for sustainable home maintenance, from choosing the right materials to adopting energy-efficient practices.

3.1 Energy-Efficient Maintenance

Maintaining energy efficiency in your home is one of the most effective ways to reduce your environmental footprint. Proper upkeep of appliances, insulation, and HVAC systems ensures that they run efficiently, using less energy and lowering your utility bills.

3.1.1 HVAC Maintenance

Your heating, ventilation, and air conditioning (HVAC) system plays a major role in your home's energy consumption. Regular maintenance ensures that it operates at peak efficiency.

- **Change Air Filters**: Dirty air filters make your HVAC system work harder, using more energy. Replace filters every 1-3 months to ensure proper airflow and efficiency.
- **Schedule Annual Tune-Ups**: Have your HVAC system serviced by a professional once a year to check for any issues and ensure it's running efficiently.
- **Seal Ducts and Windows**: Check for leaks in ductwork and around windows or doors. Sealing leaks prevents heat or cool air from escaping, reducing energy waste.

- **Programmable Thermostat**: Install a programmable thermostat to automatically adjust the temperature when you're away, saving energy without sacrificing comfort.

3.1.2 Efficient Lighting

Lighting is another area where small changes can make a big difference in your energy use.

- **Switch to LED Bulbs**: LED bulbs use up to 75% less energy and last much longer than traditional incandescent bulbs.
- **Use Natural Light**: Take advantage of natural daylight whenever possible to reduce the need for artificial lighting during the day.
- **Install Motion Sensors**: For outdoor or infrequently used indoor spaces, motion sensors ensure lights are only on when needed, saving energy.

3.2 Water Efficiency and Plumbing Maintenance

Maintaining an efficient plumbing system is key to reducing water waste in your home. By preventing leaks and using water-saving fixtures, you can conserve water and reduce your utility costs.

3.2.1 Fixing Leaks

Even small leaks can waste a significant amount of water over time. Regularly check for and fix any leaks in your plumbing system.

- **Check Faucets and Toilets**: Leaky faucets or running toilets can waste hundreds of gallons of water each month. Repair or replace faulty fixtures to stop water waste.
- **Inspect Pipes**: Look for signs of water damage, such as damp spots or mildew, which could indicate a hidden leak in your plumbing.

3.2.2 Water-Saving Fixtures

Installing water-efficient fixtures can help reduce your water consumption without sacrificing performance.

- **Low-Flow Showerheads and Faucets**: These fixtures use less water while maintaining water pressure, helping you save water during everyday tasks like showering and washing dishes.
- **Dual-Flush Toilets**: These toilets offer two flush options—one for liquid waste and one for solid waste—allowing you to use less water when possible.

3.2.3 Rainwater Harvesting

Collecting and reusing rainwater is a great way to reduce your reliance on municipal water for tasks like gardening and outdoor cleaning.

- **Install Rain Barrels**: Set up rain barrels to collect water from your gutters. This water can be used for irrigation, cleaning, or even for non-potable indoor uses if properly filtered.

3.3 Eco-Friendly Building Materials

When making repairs or upgrades to your home, choosing sustainable, eco-friendly materials can reduce your environmental impact and improve the longevity of your home.

3.3.1 Sustainable Wood and Materials

If you're undertaking renovations or repairs, consider using sustainable, responsibly sourced materials.

- **Reclaimed Wood**: Using reclaimed or recycled wood reduces the demand for new lumber and gives old materials a second life.
- **Bamboo Flooring**: Bamboo is a fast-growing, renewable resource that makes an excellent alternative to traditional hardwood flooring.
- **Recycled Insulation**: Choose insulation made from recycled materials like denim or cellulose. These options are not only eco-friendly but also highly effective at keeping your home insulated.

3.3.2 Eco-Friendly Paints and Finishes

Many conventional paints and finishes contain volatile organic compounds (VOCs), which can contribute to indoor air pollution and have harmful health effects. Opt for low-VOC or VOC-free paints and finishes to improve your home's air quality.

- **Low-VOC or No-VOC Paints**: These paints are just as effective as traditional paints but emit fewer harmful chemicals, making them a healthier choice for your home.
- **Natural Finishes**: Look for wood stains, varnishes, and finishes made from natural oils and waxes, which are less harmful to the environment.

3.4 Preventative Maintenance for Longevity

One of the greenest things you can do is to ensure that your home and appliances last as long as possible, reducing the need for replacements and the associated environmental costs. Regular maintenance can prevent breakdowns and extend the life of everything from appliances to roofing.

3.4.1 Appliance Maintenance

Regular care for your appliances ensures they run efficiently and last longer, reducing the need for replacements and the energy required to produce new ones.

- **Clean Refrigerator Coils**: Dust buildup on refrigerator coils makes the appliance work harder, increasing energy consumption. Clean the coils every few months to maintain efficiency.
- **Maintain Water Heaters**: Drain your water heater once a year to remove sediment buildup, which can decrease efficiency and shorten the lifespan of the unit.

3.4.2 Roof and Gutter Maintenance

Keeping your roof and gutters in good condition prevents water damage and ensures that your home remains well-insulated.

- **Clean Gutters**: Clogged gutters can lead to water damage, mold growth, and ice dams in winter. Clean them regularly to ensure proper drainage.
- **Inspect Your Roof**: Check for damaged or missing shingles and repair them promptly to prevent leaks and heat loss.

3.5 Sustainable Landscaping

The exterior of your home also plays a role in sustainable home maintenance. Eco-friendly landscaping practices not only enhance your home's beauty but also help conserve water, reduce pollution, and support local ecosystems.

3.5.1 Native Plants and Xeriscaping

Native plants are adapted to your local climate and soil, requiring less water and fewer chemical inputs like fertilizers and pesticides.

- **Xeriscaping**: This landscaping method focuses on using drought-tolerant plants and minimizing water use. It's an excellent option for areas with water scarcity.
- **Rain Gardens**: Planting a rain garden with native plants helps manage stormwater runoff, reduces soil erosion, and supports local wildlife.

3.5.2 Composting Yard Waste

Instead of sending yard waste like grass clippings, leaves, and branches to the landfill, composting them helps create nutrient-rich soil for your garden.

- **Composting**: Set up a compost bin or pile in your yard to turn organic yard waste into valuable compost that can be used to nourish your garden.
- **Mulching**: Use fallen leaves and other organic material as mulch to retain soil moisture, reduce weeds, and enrich the soil.

Key Takeaways

- Sustainable home maintenance is about adopting practices that reduce your environmental impact while keeping your home running efficiently and in good condition.
- Simple actions like sealing leaks, using energy-efficient appliances, and switching to eco-friendly building materials can make a big difference in conserving resources.
- Regular preventative maintenance can extend the lifespan of your home's systems and appliances, reducing the need for replacements and minimizing waste.
- Sustainable landscaping and water conservation practices further contribute to a greener home environment.

By making small adjustments in how you maintain your home, you can significantly reduce your environmental footprint while creating a healthier, more sustainable living space. In the next chapter, we'll explore eco-friendly transportation options and how you can reduce your carbon footprint on the go.

Chapter 9: Creating Green Spaces

1. Gardening for Beginners: How to Start Your Own Eco-Friendly Garden

Gardening is one of the most rewarding and accessible ways to make a positive impact on the environment right from your own home. Starting an eco-friendly garden doesn't require a lot of space or experience—whether you have a large backyard or just a small balcony, you can create a green space that supports local biodiversity, reduces your carbon footprint, and provides fresh, homegrown produce.

In this section, we'll walk you through the basics of starting your own eco-friendly garden, from planning and choosing plants to sustainable gardening practices that minimize waste and conserve resources.

9.1.1 Planning Your Garden: Where to Begin

The first step in creating an eco-friendly garden is to plan wisely. Consider the available space, local climate, and your gardening goals before planting.

Choosing the Right Location

- **Assess Sunlight**: Most plants, especially vegetables, need at least six hours of sunlight per day. Observe your space to determine where the sun shines the longest and plan to place sun-loving plants there.
- **Consider Soil**: Good soil is the foundation of a healthy garden. If you're planting in the ground, test your soil to see if it needs any adjustments (e.g., adding compost to improve texture or pH balance). If using containers, choose organic, nutrient-rich potting soil.
- **Water Access**: Ensure your garden is close to a water source, whether it's a hose, rain barrel, or watering can. This will make it easier to maintain your plants, especially during dry periods.

Decide What to Grow

Choosing what to grow depends on your space, climate, and personal preferences.

- **Start Small**: If you're new to gardening, begin with a small plot or a few containers. It's easier to manage, and you can expand as you gain experience.
- **Grow What You Eat**: Plant vegetables, fruits, and herbs that you frequently use in your meals. This not only makes your garden more practical but also reduces your need to buy these items, cutting down on food miles and packaging waste.

- **Consider Native Plants**: Native plants are adapted to your local environment, which means they require less water and are more resistant to pests. They also support local wildlife, like pollinators (bees, butterflies), which are essential for a healthy ecosystem.

9.1.2 Sustainable Gardening Practices

Once your garden is planned, adopting eco-friendly practices will help you create a truly sustainable green space. These techniques minimize resource use, promote biodiversity, and keep your garden healthy without relying on harmful chemicals.

Composting for Soil Health

Healthy soil is crucial for a productive garden. Instead of using chemical fertilizers, composting is a natural way to enrich your soil and reduce food waste.

- **Start a Compost Bin**: Composting food scraps, yard waste, and other organic materials creates nutrient-rich compost that improves soil structure and fertility. Compost bins can be as simple as a pile in your yard or a dedicated bin in a small space.
- **Use Organic Matter**: Incorporate compost into your garden beds to retain moisture, encourage earthworms, and provide nutrients to your plants.

Mulching to Conserve Water and Prevent Weeds

Mulching is an easy way to conserve water, suppress weeds, and regulate soil temperature.

- **Types of Mulch**: Organic mulches like straw, leaves, or wood chips break down over time, adding nutrients to the soil. Apply a thick layer around your plants to reduce evaporation and discourage weed growth.
- **Water Conservation**: Mulch helps keep the soil moist by reducing evaporation, which means you can water your garden less frequently, especially in hot or dry conditions.

Watering Wisely

Conserving water is a key aspect of sustainable gardening. Smart watering techniques ensure your plants get the moisture they need without wasting water.

- **Water in the Morning**: Watering early in the day reduces evaporation and gives plants time to absorb moisture before the heat of the day.

- **Drip Irrigation**: For efficient watering, consider setting up a drip irrigation system, which delivers water directly to the roots of plants. This method minimizes water waste and ensures plants get the hydration they need.
- **Rainwater Harvesting**: Collect rainwater in barrels or containers and use it to water your garden. Rainwater is free and contains natural nutrients that are beneficial for plants.

9.1.3 Organic Pest Control

A truly eco-friendly garden avoids chemical pesticides and herbicides, which can harm beneficial insects, contaminate soil and water, and disrupt local ecosystems. Instead, opt for organic, natural methods to manage pests.

Encourage Beneficial Insects

Not all insects are harmful to your garden—some are natural predators of pests or help with pollination.

- **Attract Pollinators**: Plant flowers that attract bees, butterflies, and other pollinators to help fertilize your plants. Native flowering plants like lavender, sunflowers, and coneflowers are great options.
- **Welcome Predators**: Encourage beneficial insects like ladybugs, spiders, and predatory beetles that naturally keep pest populations in check.

Natural Pest Repellents

There are many natural methods to repel pests without using harmful chemicals.

- **Neem Oil**: This natural oil, derived from the neem tree, is an effective, eco-friendly pesticide that controls aphids, spider mites, and other pests without harming beneficial insects.
- **Companion Planting**: Planting certain plants together can naturally repel pests. For example, marigolds deter aphids, and basil repels mosquitoes and flies.

9.1.4 Growing Your Own Food

One of the biggest rewards of creating a green space is growing your own food. Whether you have a small herb garden on your windowsill or a vegetable patch in your yard, growing your own produce reduces your carbon footprint and provides fresh, chemical-free food.

Choosing Easy-to-Grow Crops

If you're new to gardening, start with easy-to-grow vegetables and herbs that require minimal maintenance.

- **Herbs**: Basil, parsley, mint, and cilantro are easy to grow in small containers and can be harvested throughout the growing season.
- **Leafy Greens**: Lettuce, spinach, and kale grow quickly and can be harvested multiple times.
- **Tomatoes**: Tomatoes are a garden favorite and can be grown in the ground or in containers. Choose varieties suited to your climate for the best results.

Reducing Food Miles

Growing your own food helps reduce the distance food travels to get to your plate, lowering your carbon footprint. Plus, homegrown produce is often fresher and more flavorful than store-bought varieties, and it's free of synthetic pesticides.

9.1.5 The Benefits of Green Spaces for Mental and Physical Health

In addition to the environmental benefits, creating green spaces can improve your mental and physical health. Gardening is a relaxing, stress-relieving activity that connects you to nature, encourages physical exercise, and provides a sense of accomplishment.

- **Stress Reduction**: Spending time in your garden, surrounded by greenery, can lower stress levels and promote mindfulness.
- **Physical Exercise**: Gardening involves a range of physical activities, from digging and planting to weeding and harvesting, which helps keep you active.
- **Mental Well-Being**: Tending to a garden and watching it grow can provide a sense of purpose and satisfaction, boosting your overall well-being.

Key Takeaways

- Starting an eco-friendly garden is a great way to reduce your environmental impact, grow your own food, and create a peaceful, green space.
- Planning wisely, adopting sustainable gardening practices, and using organic methods are key to success.
- Gardening has additional benefits for your mental and physical health, making it a rewarding hobby that supports both personal well-being and environmental sustainability.

9.2 Composting Made Simple: Turning Food Waste into Nutrient-Rich Compost

Composting is one of the simplest and most effective ways to reduce waste while enriching your garden's soil. By composting, you divert organic material from landfills, reduce greenhouse gas emissions, and create a natural fertilizer that enhances soil health. Plus, it's an easy process that anyone can start, whether you have a small apartment or a large yard.

In this section, we'll break down the basics of composting, including what materials to use, how to manage your compost, and tips for troubleshooting common problems. By the end, you'll be well on your way to turning kitchen scraps into nutrient-rich compost for your garden.

9.2.1 What is Composting and Why is it Important?

Composting is the natural process of breaking down organic materials—like food scraps, yard waste, and paper—into a nutrient-dense soil amendment known as compost. This decomposition is carried out by microorganisms, worms, and insects that transform the waste into a dark, crumbly substance that can be added to your garden to boost soil health.

Why compost?

- **Reduces Waste**: Up to 30% of what we throw away is organic waste that could be composted. Composting reduces the amount of material sent to landfills, where it would otherwise generate methane, a potent greenhouse gas.
- **Enriches Soil**: Compost improves soil structure, retains moisture, and provides essential nutrients that plants need to thrive. This reduces the need for chemical fertilizers.
- **Promotes Sustainability**: Composting is a natural recycling process that closes the loop on food waste, turning what would be garbage into a valuable resource for growing more food.

9.2.2 What Can and Can't Be Composted

To get started with composting, it's important to know which materials are suitable for composting and which should be avoided. A balanced compost pile consists of two main types of materials: "greens" and "browns."

Greens (Nitrogen-Rich Materials)

"Greens" are rich in nitrogen and include moist, organic waste that breaks down quickly.

- Fruit and vegetable scraps
- Coffee grounds and filters
- Tea bags (without synthetic parts)
- Fresh grass clippings
- Plant trimmings
- Eggshells (crushed)

Browns (Carbon-Rich Materials)

"Browns" provide carbon, which is essential for balancing the compost pile and ensuring proper decomposition.

- Dry leaves
- Twigs and small branches
- Shredded paper or cardboard (avoid glossy paper)
- Straw or hay
- Wood chips or sawdust (from untreated wood)
- Paper towels or napkins (if not soiled with grease or chemicals)

What Not to Compost

There are certain materials that should be avoided in composting because they can attract pests, spread disease, or create unpleasant odors.

- Meat, fish, or dairy products
- Fats, oils, or greasy foods
- Pet waste (from dogs or cats)
- Treated wood or synthetic materials
- Diseased plants or weeds that have gone to seed

By sticking to the right balance of "greens" and "browns," you can create a healthy, efficient compost pile.

9.2.3 How to Set Up Your Compost System

There are different ways to compost, depending on the space you have and the amount of organic waste you generate. Below are some common composting methods, suitable for various living situations.

Backyard Composting

If you have a backyard, an open compost pile or a bin system is ideal for turning large amounts of organic waste into compost.

1. **Choose a Location**: Find a shady spot in your yard, preferably near a water source, where you can build your compost pile or place a compost bin. Make sure the location is accessible but not too close to your home to avoid attracting pests.
2. **Build Layers**: Start your pile with a layer of coarse materials like twigs or straw for drainage, then alternate layers of greens (food scraps) and browns (leaves, paper). A good rule of thumb is to add about two to three parts "browns" for every one part "greens."
3. **Water and Turn**: Keep the compost pile moist but not too wet. Aim for the consistency of a damp sponge. Turn the pile every few weeks to provide oxygen, which helps the microorganisms break down the materials faster.

Composting in Small Spaces

If you live in an apartment or have limited outdoor space, you can still compost using smaller, more contained systems.

- **Indoor Composting**: Composting indoors with a small bin and a carbon filter can help control odors and allows you to compost year-round. These bins can hold kitchen scraps and break them down with minimal space and effort.
- **Vermicomposting**: This method uses worms (typically red wigglers) to break down food scraps in a compact bin. Vermicomposting is ideal for small spaces since it doesn't require outdoor access, and the worms produce high-quality compost known as "worm castings" that are excellent for plants.

9.2.4 Managing Your Compost Pile

To ensure that your compost pile breaks down efficiently, you'll need to maintain a balance between greens and browns and provide the right conditions for decomposition.

Keep it Balanced

A healthy compost pile needs a good mix of nitrogen-rich "greens" and carbon-rich "browns" to maintain the right balance of nutrients.

- **Too Many Greens**: If your pile contains too many greens, it may become soggy and start to smell. To fix this, add more browns, such as dry leaves or shredded cardboard, to absorb moisture and balance the pile.

- **Too Many Browns**: If your pile has too many browns, it may take a long time to break down. In this case, add more greens, such as vegetable scraps or coffee grounds, to speed up the decomposition process.

Turning and Aerating

Turning your compost pile regularly is crucial for adding oxygen, which speeds up the decomposition process. Every few weeks, use a pitchfork or shovel to turn the pile and mix the layers. This also helps distribute moisture and heat, which encourages the microorganisms to work faster.

9.2.5 Troubleshooting Common Composting Problems

As you begin composting, you may encounter some common issues. Here are a few problems and their solutions:

- **Bad Odor**: If your compost pile smells bad, it likely has too much moisture or too many "greens." Add more dry "browns" like leaves or paper and turn the pile to add air.
- **Pile Isn't Breaking Down**: If your compost pile isn't breaking down as quickly as you'd like, it may need more nitrogen-rich "greens" or more moisture. Check that the pile is damp but not soggy, and add more greens to speed up decomposition.
- **Attracting Pests**: If you notice pests (like rodents) in your compost, you may have added meat, dairy, or oily foods. Remove these items and keep your compost covered with a lid or add a thick layer of browns to deter pests.

9.2.6 Using Your Finished Compost

After a few months, your compost will start to look dark and crumbly with an earthy smell—this means it's ready to use! Here's how to apply it:

- **In the Garden**: Mix finished compost into your garden beds to improve soil structure, boost nutrients, and encourage plant growth.
- **For Potted Plants**: Add compost to your potted plants as a top dressing or mix it with potting soil to provide extra nutrition for your indoor plants.
- **Lawn Care**: Spread a thin layer of compost on your lawn to enhance soil health and encourage lush, green grass.

By turning your food waste into compost, you create a closed-loop system that reduces waste and supports sustainable living.

Key Takeaways

- Composting is an easy and effective way to reduce waste and create nutrient-rich soil for your garden.
- Maintain a balance of "greens" (nitrogen-rich materials) and "browns" (carbon-rich materials) for optimal decomposition.
- Regularly turn and aerate your compost to speed up the process and avoid common problems like bad odors or pests.
- Use your finished compost to enrich garden soil, potting plants, or even improve your lawn, all while supporting the environment.

9.3 Growing Your Own Food: The Benefits of Growing Fruits, Vegetables, and Herbs at Home

Growing your own food is one of the most rewarding and sustainable ways to contribute to an eco-friendly lifestyle. Not only does it help reduce your carbon footprint by eliminating the need for long-distance food transportation, but it also gives you control over how your food is produced, allowing you to grow fresh, chemical-free produce. Whether you have a small balcony or a spacious backyard, you can grow a variety of fruits, vegetables, and herbs that suit your space, time, and skill level.

In this section, we will explore the many benefits of growing your own food and provide practical tips to help you get started. From saving money to enhancing your health and supporting biodiversity, growing food at home offers numerous rewards.

9.3.1 Environmental Benefits

One of the primary reasons to grow your own food is the positive impact it has on the environment. By cultivating your own garden, you're reducing your reliance on store-bought produce, which often involves significant environmental costs such as pesticide use, packaging waste, and the carbon footprint associated with transporting food across long distances.

Here's how growing your own food helps the environment:

- **Reduction in Food Miles**: The average fruit or vegetable travels thousands of miles from farm to supermarket before it reaches your plate. Growing your own food cuts down on food miles, reducing the associated greenhouse gas emissions.
- **Less Packaging Waste**: When you grow your own food, you eliminate the need for plastic packaging typically used for transporting fruits and vegetables. This helps reduce plastic waste that often ends up in landfills or oceans.

- **Pesticide-Free Food**: Homegrown produce allows you to avoid harmful pesticides and herbicides that are commonly used in conventional farming. You can opt for organic growing methods, promoting a healthier, chemical-free environment.
- **Biodiversity Promotion**: By growing a variety of plants, you support local biodiversity. A diverse garden attracts beneficial insects, pollinators like bees, and wildlife, creating a balanced ecosystem.

9.3.2 Health and Nutrition Benefits

Growing your own food can also have significant benefits for your health. When you harvest fruits and vegetables from your own garden, you're consuming fresh, nutrient-dense produce at its peak ripeness, which maximizes its nutritional value.

- **Fresher, Healthier Produce**: Homegrown food is typically fresher than store-bought produce, which can lose nutritional value over time. By harvesting food just before consumption, you ensure that your meals are packed with vitamins, minerals, and antioxidants.
- **Chemical-Free**: Growing your own food allows you to avoid pesticides, herbicides, and other chemicals commonly used in conventional farming. This is especially important for leafy greens and fruits that are often treated with harmful substances.
- **Better Flavor**: Many home gardeners find that their homegrown fruits and vegetables taste far better than store-bought varieties. The difference in flavor comes from the fact that your produce ripens naturally on the plant, without being picked early to withstand transportation.

9.3.3 Economic Benefits

Growing your own food can save you money, especially if you regularly buy fresh fruits, vegetables, and herbs. While there may be some upfront costs (such as seeds, soil, or gardening tools), the long-term savings can be substantial.

- **Lower Grocery Bills**: By growing your own food, you can significantly reduce your grocery bills, especially for items like tomatoes, lettuce, herbs, and other common vegetables. Over time, your garden will produce more than enough to meet your household's needs.
- **Long-Term Investment**: While you may need to invest in gardening supplies initially, the yields from your garden will continue to provide fresh produce year after year, making it a long-term investment.

- **Less Waste**: Growing your own food also reduces food waste. By harvesting just what you need and using fresh produce, you can minimize spoilage and avoid throwing away expired or unwanted food.

9.3.4 Mental and Physical Health Benefits

Gardening is not only a physical activity but also a therapeutic one. Growing your own food can improve your overall well-being by offering both mental and physical health benefits.

- **Stress Relief**: Gardening is known for its therapeutic effects. Spending time in nature, working with your hands, and nurturing plants can help reduce stress and anxiety, promoting relaxation and mental clarity.
- **Physical Activity**: Gardening is a moderate form of exercise that can help improve strength, flexibility, and coordination. Tasks such as digging, planting, weeding, and harvesting can burn calories and improve cardiovascular health.
- **Mindfulness**: Gardening encourages mindfulness as you connect with nature and the process of growing food. It allows you to slow down and appreciate the natural world around you.

9.3.5 How to Get Started with Growing Your Own Food

Starting your own garden doesn't have to be overwhelming. Whether you have a spacious backyard or a small balcony, you can begin growing food with just a few basic steps. Here's a guide to help you start your home garden:

1. Choose the Right Location

- **Sunlight**: Most fruits and vegetables need at least 6 hours of direct sunlight per day. Choose a spot in your yard or home that gets ample sunlight, such as a south-facing windowsill or a sunny balcony.
- **Space**: You don't need a large yard to start growing food. Even a small balcony or windowsill can accommodate containers for growing herbs, tomatoes, and salad greens.

2. Start Small

If you're new to gardening, start small with easy-to-grow plants that don't require much maintenance. Leafy greens like spinach, lettuce, and kale, as well as herbs like basil, mint, and parsley, are perfect for beginners. These plants grow quickly and are relatively low-maintenance.

3. Choose the Right Plants for Your Climate

It's important to select plants that are suited to your local climate and growing conditions. Consider the USDA Hardiness Zones and local growing seasons to choose varieties that will thrive in your area. For example, some plants may need protection during the winter, while others may require a warm, frost-free environment.

4. Grow in Containers or Raised Beds (If Space is Limited)

If you don't have access to traditional garden beds, consider using containers, pots, or raised beds to grow your food. Containers are great for small spaces, and they allow you to have full control over soil quality and watering. Raised beds also make gardening easier by reducing the need for bending and digging.

5. Plan Your Planting

Some plants do better when planted together. Companion planting involves growing complementary plants next to each other to benefit from natural pest control, improved growth, and better yields. For example, planting basil alongside tomatoes can help repel pests while boosting the flavor of the tomatoes.

6. Water and Feed Your Plants

Water your garden regularly but be mindful not to overwater, which can lead to root rot. Use organic fertilizers or compost to feed your plants and maintain soil fertility. Mulching around plants can also help retain moisture and suppress weeds.

7. Harvest and Enjoy

Once your plants are ready to harvest, enjoy the fruits of your labor! Be sure to pick produce when it's ripe to ensure the best taste and nutrient content. Many plants, such as lettuce and herbs, will continue to grow after you harvest, allowing for multiple harvests throughout the growing season.

9.3.6 Tips for Success

- **Keep Learning**: Gardening is a continuous learning process. Experiment with different plants, techniques, and gardening methods to find what works best for you.

- **Start with Easy Plants**: As a beginner, start with easy-to-grow vegetables like tomatoes, radishes, or peppers. These plants are resilient and provide fast rewards.
- **Get the Whole Family Involved**: Gardening is a great activity for families. Involve children in the process, and teach them about sustainability, healthy eating, and the importance of nature.

Key Takeaways

- Growing your own food reduces your environmental impact by cutting down on food miles, packaging waste, and pesticide use.
- Homegrown produce is fresher, healthier, and free from chemicals, offering better nutrition and taste.
- Gardening offers significant mental and physical health benefits, including stress relief and exercise.
- Starting a garden can save you money and reduce food waste in the long run.
- By starting small, choosing the right plants, and learning as you go, anyone can enjoy the rewards of growing their own food, regardless of space or experience.

In the next chapter, we'll explore eco-friendly transportation options and how switching to greener ways to travel can help reduce your carbon footprint. Stay tuned!

Chapter 10: Forming Sustainable Habits

10.1 Building Long-Term Eco-Friendly Habits

Adopting a sustainable lifestyle is about more than just making one-time changes. To make a lasting impact, it's important to build eco-friendly habits that become second nature. In this chapter, we'll explore practical strategies for turning eco-conscious actions into long-term habits, helping you stay motivated and committed to a greener way of living.

10.1.1 Understanding the Importance of Habit Formation

The key to creating lasting change lies in habit formation. Eco-friendly actions like reducing waste, conserving water, or buying sustainably often require a mindset shift. These changes don't happen overnight, but with consistent effort, they can evolve into habits that you perform automatically. This transformation is essential for maintaining an eco-friendly lifestyle in the long run.

Habits are behaviors that we repeat regularly, often without thinking. By consciously making small, sustainable choices every day, you can train your brain to adopt these actions as part of your routine. Over time, these small steps add up to significant environmental benefits.

10.1.2 Start Small and Gradually Build Momentum

When creating new habits, it's important to start small. Trying to overhaul your entire lifestyle at once can feel overwhelming and unsustainable. Instead, focus on one or two areas where you can make immediate changes. For example, you might begin by using reusable bags when shopping or cutting down on single-use plastic. Once you've incorporated these actions into your daily routine, you can move on to other habits, like reducing food waste or composting.

Starting small helps to build confidence and momentum. As you see the positive effects of your changes, you'll feel more motivated to continue and expand your efforts.

10.1.3 Track Your Progress and Celebrate Milestones

Tracking your progress is an excellent way to stay motivated. When you can see how far you've come, it boosts your confidence and reinforces the idea that your efforts are making a difference. Consider keeping a sustainability journal where you log the eco-friendly actions you take each day, such as switching off lights when not in use, reducing water usage, or cooking more plant-based meals.

Celebrating small milestones along the way can also help maintain enthusiasm. Whether it's treating yourself to something special or acknowledging your success with a friend, celebrating your achievements keeps you engaged and committed to your goals.

10.1.4 Make It Easy and Convenient

One of the best ways to build sustainable habits is to make them as easy and convenient as possible. The less effort you have to put into making a change, the more likely you are to stick with it. Here are some strategies to make your eco-friendly habits easier to implement:

- **Organize Your Space**: Keep reusable shopping bags, containers, and water bottles in easy-to-reach spots so you remember to use them. Store eco-friendly cleaning products or compost bins in areas where you'll use them most frequently.
- **Set Reminders**: Use phone apps or sticky notes to remind yourself of your sustainable goals, like turning off lights, buying secondhand, or using less water.
- **Make It Part of Your Routine**: Incorporate sustainable actions into your daily routine. For example, you might decide to walk or bike to work every day, or cook plant-based meals a few nights each week. When these actions become part of your routine, they feel natural.

10.1.5 Be Flexible and Adaptable

Life can be unpredictable, and sometimes eco-friendly habits can be disrupted by external factors like a busy schedule, travel, or unexpected events. It's important to be flexible and adaptable. If you can't compost while traveling or forget your reusable shopping bags on occasion, don't be too hard on yourself. The key is to keep going and get back on track as soon as possible. Developing eco-friendly habits is a long-term journey, and occasional setbacks are part of the process.

10.1.6 Find Accountability and Support

Having support can be one of the most powerful ways to stick with new habits. Whether it's a friend, family member, or online community, having someone to share your journey with can keep you accountable and motivated. Here are some ways to find support:

- **Join Eco-Friendly Communities**: Many online forums, social media groups, or local organizations offer resources and encouragement for sustainable living. Connecting with others who share your values can help you stay inspired and exchange ideas.
- **Set Joint Goals**: If you live with others, like a partner or family, set sustainability goals together. Having a collective goal can increase motivation and foster a sense of teamwork.
- **Engage in Eco-Challenges**: Participate in eco-challenges, such as a 'Plastic-Free July" or "Meatless Mondays," to test your ability to stick with new habits in a fun and supportive way.

10.1.7 Keep Educating Yourself and Stay Inspired

The more you learn about environmental issues and solutions, the more motivated you'll feel to take action. Keep reading about sustainability, watch documentaries, or attend local events related to environmental conservation. Staying informed helps you understand the impact of your choices and gives you new ideas to incorporate into your life.

It's also important to stay inspired by the positive changes you're making. Whether it's reducing waste, eating more plant-based foods, or conserving energy, remember that every small action counts. Your commitment to sustainable living is part of a global effort to create a healthier planet.

10.1.8 Embrace the Process, Not Just the Outcome

Lastly, remember that adopting a more sustainable lifestyle is a journey, not a destination. Rather than focusing solely on achieving perfection or reaching specific goals, embrace the process of making mindful, eco-friendly decisions each day. The more you make these choices a natural part of your life, the more fulfilling the experience will be.

10.1.9 Key Takeaways

- **Start small and build momentum** by making manageable changes to your lifestyle.
- **Track your progress** and celebrate milestones to stay motivated and encouraged.
- **Make eco-friendly habits easy and convenient** by organizing your environment and incorporating them into your routine.

- **Be flexible** when life's challenges disrupt your habits, and remember that setbacks are part of the process.
- **Find support** from friends, family, or online communities to help you stay accountable and motivated.
- **Keep learning** about sustainability to stay inspired and informed about new ways to reduce your environmental impact.
- **Embrace the process** of adopting sustainable habits, and remember that every small change counts.

By building sustainable habits, you'll not only reduce your environmental impact but also improve your overall quality of life. As you continue to grow and learn, you'll see the positive changes that come from living in harmony with the planet. Stay committed, stay inspired, and enjoy the journey toward a greener, more sustainable future.

10.2 Involving Family and Friends: How to Inspire Others to Join Your Green Journey

One of the most rewarding aspects of living sustainably is sharing your journey with others and inspiring them to take action too. The more people who adopt eco-friendly habits, the greater the collective impact on the environment. In this section, we'll explore how you can engage your family, friends, and community to join you in making greener choices.

10.2.1 Lead by Example

The most powerful way to inspire others to join your green journey is by leading through your actions. When your family and friends see you consistently making eco-friendly choices, they are more likely to be curious about how they can do the same. Lead by example in small, everyday actions such as:

- **Reducing Waste**: Show how easy it is to refuse single-use plastic, compost food scraps, or recycle properly. When others see how much waste you're diverting from landfills, they may want to adopt similar habits.
- **Energy Efficiency**: Demonstrate how simple steps like turning off lights when not in use, unplugging electronics, or switching to LED bulbs can save energy and reduce your carbon footprint.
- **Eco-Friendly Transportation**: Walk, bike, carpool, or use public transportation whenever possible, and talk about the environmental benefits of reducing car travel.

By living in alignment with your values, you set a natural example for others to follow.

10.2.2 Start Conversations About Sustainability

Sometimes, people aren't aware of the environmental benefits of certain actions until they are introduced to them. Starting conversations about sustainability with your family and friends can plant the seed for change. Here are a few ideas to help open the discussion:

- **Share Articles or Documentaries**: If you come across an interesting article, video, or documentary about sustainability, share it with your loved ones. It could be about climate change, waste reduction, or the importance of eating locally grown food. These resources can help raise awareness and spark curiosity.
- **Highlight the Benefits**: Emphasize how adopting eco-friendly habits can improve health, save money, and even make life easier. For example, walking or biking instead of driving reduces fuel costs and provides more exercise. Composting can reduce waste while enriching your garden soil.
- **Use Positive Language**: Frame sustainability as a positive choice rather than a sacrifice. Rather than focusing on the things you "can't" do, talk about the exciting new habits you've gained—like trying new plant-based recipes or discovering eco-friendly cleaning products.

10.2.3 Involve Children in Eco-Friendly Activities

Children are often eager to learn and take part in hands-on activities, which makes them ideal partners for engaging in green initiatives. By involving your kids or younger relatives in eco-friendly activities, you not only teach them valuable skills but also create a sense of shared responsibility for the planet. Here are some fun and educational activities to try:

- **Gardening Together**: Planting a garden with children is a great way to teach them about where food comes from and the importance of growing sustainably. They can learn about composting, soil health, and the benefits of growing their own food.
- **Eco-Crafts and Upcycling**: Show children how to repurpose materials around the house into new, useful items. Making art from recycled materials is not only fun, but it's also an opportunity to discuss waste reduction and the importance of reusing.
- **Nature Walks and Cleanups**: Take the family on nature walks, clean up litter, or organize neighborhood clean-up days. This helps instill a

respect for nature and a sense of environmental stewardship in young people.

When you involve children, you foster a mindset that values sustainability, and they are more likely to carry these lessons into adulthood.

10.2.4 Make it Social and Fun

Changing habits can sometimes feel like a chore, but when you make it social and fun, it becomes more enjoyable and engaging. Consider organizing eco-friendly events with your family and friends that allow everyone to participate and learn together:

- **Eco-Friendly Potlucks**: Host a sustainable potluck where everyone brings a dish made with local, seasonal ingredients. It's a great way to introduce others to eco-friendly eating habits while sharing delicious food.
- **Sustainability Challenges**: Set up friendly challenges with your friends or family. For example, you could have a "Plastic-Free Week," where everyone tries to avoid using plastic packaging. Or challenge each other to see who can come up with the most creative way to reuse or repurpose an item.
- **Green DIY Projects**: Get together for a DIY craft or home improvement project that promotes sustainability, such as making homemade cleaning supplies, building compost bins, or creating an indoor herb garden.

Making sustainability a fun, communal activity will encourage others to embrace it as part of their everyday lives.

10.2.5 Offer Support and Encouragement

Adopting eco-friendly habits can be difficult at first, so offering support and encouragement to those trying to make the change is crucial. Here are some ways you can help:

- **Be Patient and Non-Judgmental**: Understand that everyone is on their own journey, and it's okay if others don't adopt every habit right away. Offer encouragement and celebrate small victories, whether it's switching to reusable bags or reducing food waste.
- **Provide Resources**: Share information, tips, and resources that make it easier for others to get started. Whether it's a guide to zero-waste

living or a list of eco-friendly brands, providing practical tools can help others make informed choices.
- **Offer Help with Challenges**: If someone is struggling to make a change, offer to help them find a solution. For instance, if a friend wants to reduce plastic waste but isn't sure where to start, help them find alternatives like reusable food wraps, containers, or products from sustainable brands.

By offering ongoing support, you create an environment where others feel empowered to make and sustain eco-friendly changes.

10.2.6 Encourage Collective Action

One of the most impactful ways to inspire others is by encouraging collective action. When people see that sustainability is a group effort, it can feel more achievable and powerful. Here are a few ways to encourage collective eco-friendly efforts:

- **Organize Community Clean-Ups or Events**: Team up with local organizations to organize clean-up days, tree-planting events, or sustainability workshops. Working together toward a common goal helps strengthen the sense of community while making a positive environmental impact.
- **Form Green Support Circles**: Create a group with friends or neighbors where you can meet regularly to share eco-friendly tips, set sustainability goals, and track progress. The support of a community can help everyone stay motivated and inspired.
- **Advocate for Sustainability**: Encourage your family and friends to get involved in environmental advocacy. This could include supporting eco-friendly policies, writing to local representatives about sustainability issues, or participating in global movements like Earth Day.

When people work together, the impact is amplified, and the sense of collective accomplishment strengthens the commitment to green living.

10.2.7 Key Takeaways

- **Lead by example**: Your actions can inspire others to make eco-friendly choices.
- **Start conversations about sustainability** by sharing resources and highlighting the benefits of greener habits.

- **Involve children in eco-friendly activities**, teaching them the value of sustainability from an early age.
- **Make it social and fun** by organizing events or challenges that promote sustainability.
- **Offer support and encouragement** to family and friends as they begin their own eco-friendly journey.
- **Encourage collective action** by getting involved in community sustainability projects and advocating for change.

By involving your family, friends, and community, you create a ripple effect that helps spread sustainability and environmental awareness, making a larger impact on the planet. When everyone works together, the journey to a greener future becomes even more rewarding and effective.

10.3 Dealing with Eco-Anxiety: How to Stay Positive and Avoid Feeling Overwhelmed by Environmental Issues

As we become more aware of the environmental challenges facing our planet, it's natural to feel a sense of urgency and even anxiety about the future. From climate change to pollution, the scale of environmental issues can feel overwhelming at times. However, while it's important to acknowledge the gravity of the situation, it's equally crucial to find ways to manage eco-anxiety so that it doesn't paralyze you or prevent you from taking action. In this section, we'll explore how to cope with the anxiety that comes with environmental awareness and focus on constructive steps you can take to maintain a positive outlook while contributing to a better future.

10.3.1 Understand That Eco-Anxiety is Normal

Eco-anxiety is a natural response to the overwhelming evidence of environmental degradation, and you're not alone in feeling this way. As global challenges like climate change, deforestation, and biodiversity loss gain more attention, many people experience stress, fear, or even guilt. These feelings are not a sign of weakness but a reflection of our deep connection to the planet and our concern for its future. Recognizing that eco-anxiety is a shared experience can help you feel less isolated and more empowered to manage these emotions.

10.3.2 Focus on What You Can Control

While it's true that the state of the planet can sometimes feel beyond our individual control, focusing on the positive steps you can take helps redirect your energy into productive actions. It's important to remember that every small change you make has an impact, and collective action, even on a small scale, can lead to significant transformation.

Here are some ways to focus on what you can control:

- **Start with your own actions**: Take pride in the eco-friendly choices you make each day, whether it's reducing waste, conserving energy, or supporting sustainable businesses. Each of these actions contributes to the larger environmental movement.
- **Engage in local sustainability efforts**: Join community groups, participate in clean-up events, or advocate for local environmental policies. These smaller-scale actions help foster a sense of agency and community involvement, which can reduce feelings of helplessness.
- **Support companies and initiatives you believe in**: When you support businesses that align with your values, you contribute to the growth of the green economy. Voting with your wallet can have a powerful ripple effect, encouraging more businesses to adopt sustainable practices.

By taking proactive steps, you can feel empowered and remind yourself that your actions, no matter how small, do make a difference.

10.3.3 Break Down Big Issues into Manageable Steps

The environmental crisis can feel massive, and it's easy to become overwhelmed when thinking about the scale of the issues. Instead of viewing them as one insurmountable challenge, try breaking them down into smaller, more manageable steps. This approach helps reduce feelings of helplessness and allows you to focus on tangible solutions.

For example:

- **Climate Change**: Focus on reducing your personal carbon footprint through transportation choices, energy efficiency, and sustainable consumption. Every reduction in carbon emissions counts.
- **Biodiversity Loss**: Support wildlife conservation efforts, plant native species in your garden, or reduce your use of pesticides. Even small actions like these contribute to the preservation of ecosystems.
- **Plastic Pollution**: Reduce your use of single-use plastics, buy in bulk, and recycle more. Every piece of plastic avoided or properly recycled adds up.

Focusing on specific areas allows you to direct your energy toward actions that can truly make a difference, giving you a sense of accomplishment and purpose.

10.3.4 Connect with Like-Minded People

One of the best ways to manage eco-anxiety is by connecting with others who share your commitment to sustainability. Surrounding yourself with like-minded people creates a support network that can help you feel less alone in your journey and give you motivation to keep going, even when things seem bleak.

Here are some ways to connect with others:

- **Join online communities**: Many social media platforms have groups focused on sustainability and environmental issues. Join forums or Facebook groups where you can share ideas, solutions, and successes with others who are equally passionate.
- **Attend local sustainability events**: Whether it's a clean-up day, a climate change protest, or a workshop on eco-friendly living, attending local events helps you connect with your community and find a sense of solidarity.
- **Start your own group**: If you don't already have a group of like-minded individuals nearby, consider starting one. Whether it's a book club focused on environmental topics or a neighborhood gardening initiative, creating a space for others to gather and share can be incredibly rewarding.

Building a supportive community fosters optimism and collective action, and it reminds you that you're not alone in your efforts.

10.3.5 Practice Self-Care

Taking care of your mental and emotional well-being is essential when dealing with eco-anxiety. If you allow yourself to become too consumed by environmental concerns, it can lead to burnout, stress, and exhaustion. Self-care is crucial in maintaining balance and ensuring that you can continue making a positive impact.

Here are some self-care practices to consider:

- **Disconnect from the news**: Constantly consuming negative news about the environment can intensify anxiety. Set aside time to unplug

from the media and focus on other activities that bring you peace, such as reading, exercising, or spending time in nature.
- **Engage in mindfulness and relaxation**: Practices like meditation, deep breathing, or yoga can help reduce stress and anxiety. Mindfulness allows you to stay present and focused on the actions you can take today, without being overwhelmed by the future.
- **Spend time in nature**: Immersing yourself in nature not only provides a sense of peace but also reinforces why protecting the planet matters. Whether it's a walk in the park, a hike in the mountains, or simply sitting outside, connecting with nature can help restore your sense of purpose and calm.

By taking care of yourself, you build the resilience needed to continue your green journey and avoid becoming overwhelmed by the challenges ahead.

10.3.6 Find Hope in Action

Ultimately, one of the best antidotes to eco-anxiety is to take action. While the scale of the environmental issues we face can seem daunting, the collective power of small, positive actions can lead to real change. Focus on the progress that's being made, whether it's in renewable energy, waste reduction, or policy reforms. Every step forward, no matter how small, is a victory.

By shifting your mindset from one of despair to one of hope and action, you'll not only help reduce your eco-anxiety, but you'll also inspire others to join in and take action. The more people who work together toward sustainable goals, the closer we get to a healthier planet for all.

10.3.7 Key Takeaways

- **Eco-anxiety is normal** and reflects your concern for the planet. Acknowledge it and recognize that you're not alone in feeling this way.
- **Focus on what you can control** by taking small, positive actions in your own life and engaging in local sustainability efforts.
- **Break down big environmental issues** into manageable steps, and focus on specific actions you can take to address them.
- **Connect with like-minded people** for support, encouragement, and collective action.
- **Practice self-care** to maintain emotional and mental well-being, so that you can continue your journey without burning out.
- **Find hope in action** by focusing on the progress being made and staying motivated by the positive impact of your contributions.

By managing eco-anxiety and focusing on constructive actions, you can maintain a positive outlook, feel empowered, and stay motivated to continue making a difference in the world.

Conclusion: Your Green Journey

As you reach the end of this book, it's time to reflect on the journey you've taken toward living a more eco-friendly life. Adopting sustainable habits and embracing a greener lifestyle is a long-term commitment, but every step you've taken brings us closer to a more sustainable world. This conclusion is about celebrating the progress you've made and looking forward to the future with a renewed sense of purpose and determination.

1.1 Reflecting on Your Progress

Think about the changes you've made so far, no matter how big or small. Perhaps you've reduced your waste, switched to renewable energy, started buying locally produced food, or begun using reusable products instead of single-use plastics. Maybe you've even inspired friends and family to follow your lead and adopt some of these practices themselves.

It's important to recognize the value of these actions. You may not yet have transformed every part of your life, but each eco-friendly choice you make sends a powerful message—not only to others, but also to yourself. You're living with intention, striving to make a difference, and helping to create a more sustainable world. Remember, every step counts, and progress is not always linear. Some days will be easier than others, and setbacks are part of the journey. What matters most is that you continue, no matter how small the steps may seem.

1.2 Looking Ahead: Setting Future Goals

Your green journey doesn't stop here. The road ahead is filled with endless opportunities to make positive changes, both for yourself and for the planet. Here are some ways to keep moving forward:

- **Set New Goals**: As you continue to build on the habits you've adopted, consider setting new, more ambitious goals. For example, you could aim to reduce your carbon footprint even further, start composting at home, or engage in local environmental initiatives. Make sure your goals are specific, achievable, and aligned with your values.
- **Stay Educated**: The world of sustainability is constantly evolving. Stay curious and keep learning about new ways to live more sustainably. Attend workshops, read books, follow environmental blogs, and stay informed about policy changes that can affect your lifestyle. The more you know, the better equipped you'll be to make informed decisions.

- **Inspire Others**: As you continue to make changes in your own life, use your experiences to inspire others. Share your successes, challenges, and the reasons behind your choices. By talking about your green journey, you can encourage friends, family, and colleagues to consider their own environmental impact and take action.
- **Track Your Progress**: Regularly check in on your sustainability goals to see how far you've come. Tracking your progress can be motivating and help you stay on track. Consider using a journal or an app to record your actions and reflect on how your habits are evolving over time.

1.3 Embracing the Future

The path to sustainability is ongoing and full of learning, adaptation, and growth. The decisions you make today are shaping the world of tomorrow. By committing to a green lifestyle, you're becoming part of a global movement that is pushing for positive environmental change, a healthier planet, and a brighter future for generations to come.

Even if the challenges seem large, remember that the power to create change lies in our collective efforts. Each individual action, no matter how small, contributes to the larger goal of a more sustainable, eco-friendly world. You're not just an individual; you are part of a community of changemakers who are shaping a better future.

1.4 Key Takeaways

- **Celebrate your progress**: Reflect on the positive changes you've already made and take pride in your commitment to living more sustainably.
- **Set new, realistic goals**: Challenge yourself to continue growing, whether through new habits or bigger sustainability projects.
- **Stay curious and engaged**: The journey doesn't end here. Continue learning, stay informed, and adapt to new sustainability practices as they arise.
- **Inspire others**: Share your knowledge, experiences, and goals with friends and family to build a network of eco-conscious individuals.
- **Keep moving forward**: Every step you take, no matter how small, makes a difference. Stay motivated and keep striving for positive environmental impact.

As you reflect on your journey and plan for the future, remember that every effort, no matter how small, contributes to a bigger change. Your green journey is just beginning, and the impact of your actions will ripple outward, inspiring others and contributing to a more sustainable future for all. Keep moving forward with hope, purpose, and the knowledge that you are making a difference.

What's Next?

As you conclude this book, you're not just finishing a chapter—you're embarking on a lifelong journey toward sustainability and environmental stewardship. Your commitment to living an eco-friendly life doesn't stop here; instead, it's a continuous process of learning, growing, and adapting. The world of sustainable living is constantly evolving, and there's always more to discover. By continuing to engage with the resources around you and staying open to new ideas, you'll find more ways to deepen your green journey and further reduce your impact on the planet.

2.1 Continuing to Learn and Grow in Your Eco-Friendly Lifestyle

Sustainability is not a destination but a path. The more you explore, the more you'll uncover about the vastness of environmental challenges and the creative solutions available. Here's how to continue building on the foundation you've laid:

1. **Keep Educating Yourself**: The world of sustainable living is vast, and there's always more to learn. Continue to read books, follow sustainability blogs, listen to podcasts, and watch documentaries that provide fresh perspectives on eco-friendly living. The more you know, the more empowered you'll feel to take meaningful action.
2. **Explore New Sustainability Practices**: Each season, explore new ways to make your life more sustainable. For instance, in the summer, you might explore growing your own herbs, and in winter, you might focus on energy-saving measures. The key is to never stop exploring and experimenting with new habits.
3. **Celebrate Your Wins and Learn from Challenges**: As you work toward a greener lifestyle, take time to acknowledge the wins—whether it's reducing your waste, eating more plant-based foods, or getting involved in local environmental activism. Equally important is reflecting on your challenges: What didn't work? Why? Use these experiences to refine your approach and improve your habits.
4. **Stay Connected with the Eco-Community**: Sustainability is a movement, and there's power in numbers. Stay connected with others on the same path—whether through social media, local groups, or community events. Engaging with others helps you stay motivated, share ideas, and build a collective impact.

5. **Set Long-Term Goals**: Don't be afraid to dream big. What do you hope to achieve in the next 1, 5, or 10 years? Whether it's achieving zero-waste status, making your home energy self-sufficient, or being a voice for environmental change, set clear, achievable goals that will keep you moving forward.

As you continue your green journey, remember that every action, no matter how small, is part of a much larger collective effort to protect our planet. Stay curious, stay committed, and keep growing!

Appendices

To help you on your path to a greener life, this section provides additional resources, guides, and recommendations.

Eco-Friendly Product Recommendations

Making sustainable choices extends beyond habits—it also involves the products we purchase. Here is a list of eco-friendly brands and products that align with a greener lifestyle:

1. **Eco-Friendly Cleaning Supplies**
    - *Method*: Non-toxic and biodegradable cleaning products.
    - *Seventh Generation*: Offers a wide range of plant-based cleaning solutions.
    - *Blueland*: Reusable cleaning bottles with concentrated refill tablets to reduce plastic waste.
2. **Sustainable Clothing**
    - *Patagonia*: Known for its environmental advocacy and high-quality, eco-friendly clothing.
    - *Reformation*: Focuses on ethical sourcing and sustainable manufacturing processes.
    - *Everlane*: Transparent sourcing with a focus on using sustainable materials.
3. **Zero-Waste Products**
    - *Hydroflask*: Reusable, durable water bottles to reduce single-use plastic.
    - *Bee's Wrap*: Sustainable, reusable food storage wraps made from organic cotton, beeswax, and tree resin.
    - *Stasher*: Silicone bags for storage, offering an alternative to plastic bags.
4. **Sustainable Food and Packaging**

- *Thrive Market*: Organic food and eco-friendly products delivered to your door.
- *Bulk Barn*: Buy food and household items in bulk to reduce packaging waste.
- *Abigo*: Eco-friendly food packaging alternatives.

5. **Energy-Saving Products**
    - *Philips Hue*: Energy-efficient smart bulbs that can be controlled to reduce electricity usage.
    - *Nest Thermostat*: Helps you manage your home's energy consumption, optimizing heating and cooling.

Each of these products and brands can help reduce your environmental footprint while providing high-quality, sustainable options for your home, wardrobe, and lifestyle.

Checklists for Green Living

Staying on track with your eco-friendly habits is easier with the right tools. Below are printable checklists to help you continue your sustainable journey:

1. **Daily Eco-Friendly Habits Checklist:**
    - Reduce water use (turn off faucets when not in use, use a low-flow showerhead)
    - Choose reusable items (bags, bottles, containers)
    - Avoid single-use plastics (straws, cups, cutlery)
    - Sort waste for recycling or composting
    - Make sustainable food choices (buy locally, choose plant-based options)
    - Save energy (unplug electronics, turn off lights)
2. **Weekly Eco-Friendly Tasks Checklist:**
    - Meal prep using fresh, seasonal ingredients
    - Plan low-impact transportation (walk, bike, carpool)
    - Wash clothes in cold water, air dry when possible
    - Shop for eco-friendly products (from local markets or bulk bins)
    - Review energy consumption (check thermostat settings, use energy-efficient appliances)
3. **Monthly Green Living Goal Tracker:**
    - Set a new sustainability goal (reduce plastic usage, increase recycling efforts, etc.)
    - Assess progress (how much waste have you reduced? How much energy have you saved?)

- Celebrate successes (reward yourself for achieving milestones)
- Reflect on challenges and adjust your approach

By following these checklists and revisiting them regularly, you'll stay on track and continue to evolve toward a more sustainable lifestyle.

In conclusion, your green journey is a lifelong adventure. There's no finish line, only ongoing opportunities to learn, grow, and make a lasting impact on the planet. By staying committed, informed, and inspired, you will continue to lead a greener, more sustainable life, helping protect the earth for generations to come.

www.ingramcontent.ccm/pod-product-compliance
Lightning Source LLC
Chambersburg PA
CBHW050001230526
45465CB00003BB/1215